VODKA

This edition published in 2015

Abridged from *Spirits & Cocktails* (1998) and
The Complete Bartender's Guide (2013)

Copyright © Carlton Books Limited 1998, 2013, 2015

Carlton Books Limited
20 Mortimer Street
London W1T 3JW

A CIP catalogue record for this book is available from the British Library

ISBN: 978-1-85375-786-0

Editor: Martin Corteel
Designer: Luke Griffin
Picture research: Paul Langan
Production: Rachel Burgess

Printed in China

MEASUREMENT CONVERSIONS

(Millilitres to Fluid ounces (UK))	
	60 ml = 2 fl oz
	75 ml = 2⅔ fl oz
	80 ml = 2¾ fl oz
10 ml = ⅓ fl oz	90 ml = 3¼ fl oz
20 ml = ⅔ fl oz	100 ml = 3½ fl oz
30 ml = 1 fl oz	125 ml = 4⅓ fl oz
40 ml = 1⅓ fl oz	150 ml = 5¼ fl oz
50 ml = 1¾ fl oz	200 ml = 7 fl oz

VODKA

DAVE BROOM

PRION

CONTENTS

INTRODUCTION

Vodka has enjoyed a rich and varied life. It's been drunk just about everywhere from the great houses and palaces of Poland and Russia to shamen's altars in Siberia, not to mention every bar in the Western world. Yet how much do we actually know about this spirit? Vodka may play a part in the daily ritual in Eastern Europe, but until quite recently few in the West give it a second thought.

Vodka, you would think, is vodka. It has made its name by being the perfect partner for any mixer – flavourless, but with an alcoholic kick, undemanding and malleable. It's ideally suited to a lifestyle where easy choices are the norm, where you don't have to try too hard. Why grapple with the intricacies of Proust when you can be "entertained" by Jeffrey Archer? Vodka's been a bit like that.

But look again … things are changing. A few years ago, if you walked into a bar and wanted a vodka it would be poured out of the well; if you were lucky, you might have got Smirnoff Blue. Then Absolut arrived and the world was

never quite the same again; it made vodka become hip, rather than just popular. It fitted the 1980s like a glove. It was clean, pristine, designed, the drinks world's equivalent of a Donna Karan suit.

But, despite the fact that Absolut has been superbly marketed (and is a fine vodka), its success was underpinned by the same principle that had taken vodka from foreign oddity to one of the world's most popular spirits – namely, vodka tasted of, well, nothing. People were, and still are, being asked to pay large amounts of money for something that apparently claims to be flavourless. It's this paradox that lies at vodka's cool heart.

THE ESSENCE OF THE SPIRIT

"I began to think vodka was my drink at last ... it went straight down into my stomach like a sword swallower's sword and made me feel powerful and godlike."

Sylvia Plath

When you look at the subject, you find that true vodka is anything but tasteless; it has verve and subtlety. The best of the premium vodkas make you look differently at spirits – their only equivalent is, perhaps, the new wave of grappas that are coming out of Italy. To get to grips with quality vodka requires you to open your mind to whispers of aroma, nuances of flavour. But this is miles from the image that most people have of this spirit. Only by looking at vodka's convoluted history are you able to get to grips with how this refined spirit has acquired this rather schizoid image.

Bill Samuels of Maker's Mark is fond of saying that bourbon died when Smirnoff started to be distilled in America and, to understand how vodka became the great mixable spirit, it's Smirnoff's story that you have to follow. What the brand did, brilliantly, was to take a drink redolent with negative imagery – hard drinking, strange ritual and, remember, this is during the Cold War – and make a household brand. Smirnoff's early US advertising claimed it was "the drink that leaves you breathless", and it certainly did that in the way that it quickly scaled the heights of the US market.

Vladimir Smirnov (sic) had fled his homeland after the Russian Revolution, and then tried to establish Smirnov distilleries in Constantinople, Lvov and Paris (see below). However, by the time he got to Paris, like so many other Russian emigrants, he was broke. Enter Rudolph Kukhesh, an ex-supplier of alcohol to the Smirnovs who had moved to the USA, changed his name to Kunett and started working for Helena Rubinstein's cosmetics company. In Paris on business, he met Smirnov who, by now down on his luck, gave him the rights and licence to sell the Smirnov portfolio in North America. Kunett changed the firm's name to Pierre Smirnoff & Fils, and started distilling in America in March 1934. It wasn't the runaway success he had hoped for. Vodka had been drunk during Prohibition, but it was one of many spirits which emerged from that period with a battered reputation – not surprisingly, since the bulk of the "vodka" that would have been sold was made in bathtubs in the back streets of Chicago and New York.

ABOVE: *A bottle of Krolewska Vodka from Poland*

In fact in 1937, Kunett, close to bankruptcy, sold the Smirnoff licence to John Martin, president of Heublein, who nearly lost his job as a result. That's how poor an image vodka had. Heublein had to wait until after the Second World War for its troublesome new charge to prove its worth. Allied to some clever marketing, vodka became the spirit that the post-war market wanted. It mixed happily, it was light, it was un-demanding. It was so versatile that barmen could create a huge number of new recipes for cocktails, but it was one of the first spirits that allowed you to make cocktails at home – add ginger beer and you had a Moscow Mule, a screwdriver was Smirnoff and orange juice. No matter what you threw at vodka, it accepted it quite happily. It even made Martinis!

Vodka was sophisticated, easy, fun – and it didn't make your breath smell. Within a brief time it had evolved into a different drink from the one which is still consumed in Eastern Europe.

Smirnoff began distilling in the UK, Canada, Australia, New Zealand and across America. Other vodkas followed in its wake – brands that were bland, neutral alcoholic bases for soft drinks and mixers.

By 1975, in terms of sales, vodka had become America's most popular spirit, and Smirnoff and Bacardi began slugging it out as the world's biggest spirit brand. The triumph of the neutral white spirit was complete. Russia's "little water" (vodka's origi-nal meaning) had grown up.

But while Bacardi had become a category in itself, Smirnoff, though a massive brand, couldn't do the same. Vodka was the winner, but rather than being tied to one particular brand remained rather anonymous. You may have got a Smirnoff in a bar, but you wouldn't have asked for it by name.

This situation existed until Absolut swung into town with a marketing campaign that others have tried desperately to imi-tate but will never replicate. Absolut was right for its time. It tied itself to cutting-edge fashion and modern art. It was ir-reverent, weird, wacky, but it was never cheap. Rather, it was elitist and utterly, utterly hip. Absolut, thanks to its success in educating barmen and wooing design-conscious consumers, began to be asked for by name. The way was now open for people to look at vodka in a different light.

ABOVE: *The impeccable
style of Absolut*

Absolut managed to bridge a gap that Finlandia and Stolichnaya had already been working on – though without much success. It was a vodka that could be used for cocktails (and push the price of the cocktail up), but it was also promoted to be drunk as a shooter – which was something that no one would ever have thought of doing with Smirnoff. New opportunities suddenly appeared for vodka.

Then, the Iron Curtain fell. Now, finally, there was a chance for Stolichnaya, Moskovskaya, Polish vodkas and the classic flavoured styles, which had been made for centuries, to make their mark in the world.

The simple reason that these vodkas are now the driving force in the market is that "premium" equates with "imported" – whether that is Polish, Russian, Finnish, Swedish or Danish. By hailing from a vodka-producing country, these brands appeal to the new consumer who wants to drink spirits that are somehow "authentic".

In the USA, this new image has now been widened to include domestic brands, which have cunningly put a different spin on the taste of nothing by emphasizing production and quality. These days you'll find pot-still vodka from Texas, organic vodka from Kentucky, vodka from glacier water from the Tetons. So successful has this shift upmarket been that, to grab a slice of this lucrative sector, Smirnoff has gone back to Moscow and started to produce a vodka from pot stills.

These new premiums are drunk neat or contribute to the revival of the Martini (though being a bluff old traditionalist, I still believe a Martini is made from gin – no other spirit will do).

Just as with any quality spirit, the vodka drinker wants authenticity – although that term is a highly personal one. It means that "real" vodka has finally got the chance to show its true colours. But what is "real" vodka?

MAKING VODKA

"I never have more than one drink before dinner. But I do like that one to be large and very strong and very cold and very well made."

James Bond, *Casino Royale*

In technical terms, vodka is pure (usually rectified) spirit that has been diluted with water and filtered before bottling. The aim has always been to look for the purest spirit possible. But why did Poland and Russia decide that a flavoursome grain spirit – akin to the early whiskies made in Scotland and Ireland – wasn't for them? One answer was the need to re-distil the spirit from the early log stills; another is down to the climate. Low-alcohol spirits freeze. If you were wanting to transport spirit during the bitter winters, it made sense to have as high-strength a spirit as possible – and that meant redistillation.

In addition, it is worth remembering that even in the early days vodka was flavoured, and that while herbs were used as masking agents, they were also there to be tasted

(the complexity of the recipes is evidence of this). The need, therefore, was for as light a spirit base as possible.

Neutral spirit can be made from anything that contains starch – in principle, vodka can be made from molasses, sugar beet, potatoes, rye, wheat, millet, maize, whey and even rectified wood alcohol.

Most basic commercial brands these days will use molasses, but premium vodkas need to retain the finer qualities of their raw material – and the best are made from either grain or potato. "Vodka seems to be the simplest liquor," says Dr Boleslaw Skrzypczak, one of Poland's recognized vodka experts. "But this understanding only skims the surface. In reality a host of factors influence the quality of vodka – proper raw material and technology for producing raw spirit, the effectiveness of spirit purification and water quality."

RAW MATERIALS

The early distillers made their spirit from the most widely available source of starch. That meant wheat in Sweden, rye and potatoes in Poland, rye and wheat in Russia. Already you are looking at different styles. Of the widely available premium vodkas today, Absolut and Altai are made from winter wheat, Moskovskaya is a classic Moscow rye (although Stolichnaya also uses wheat), while Luksusowa and Chopin are potato spirits. Rye gives bite and weight, wheat a delicacy, while potato gives a distinctive creaminess to the spirit.

The mention of the last ingredient is always liable to produce a strange reaction from Westerners. Potatoes are seen as a sign of an inferior spirit, something rustic and crude. You use potatoes to make hooch in prison, not in a distillery. While golden fields of grain are infinitely more alluring (and photograph better), a vodka like Luksusowa proves that potatoes can result in a beautifully rich, creamy spirit that's more than a match for grain vodka. In Poland, only special high-starch varieties grown along the Vistula river and on the Baltic coast are used for vodka production – and although they give less alcohol than wheat, they are still preferred for the distinctive character of the final spirit. Dr Skrzypczak also claims that, since potatoes have fewer aromatic compounds, they are better for producing neutral rectified spirits.

WATER

One distiller claims that 60 per cent of vodka's quality is down to the water used – although quite how this figure is quantified isn't specified. Water, however, is of vital importance to any spirit, and vodka is no exception. Expert tasters agree that Moscow vodka's quality suffered when the water supply changed. Finlandia, rightly, can point to its pure water source as one of the major contributory factors in its natural clean taste, as can the Siberian brand Altai, while Absolut has its own well.

Water is used twice – once for mashing, and then again at the end of the process when the spirit is diluted prior to filtration. While some distilleries can use pure spring water at this stage, others have to soften and demineralize the water in order to prevent any clouding of the spirit. Distilled water was widely used, but it's agreed that it gives a flat flavour to the spirit and, for quality brands at least, is not used. Nothing must get in the way of the pure flavour.

DISTILLATION

That purity has been achieved by a highly controlled distillation process. Vodka distillers will point out that after ferment they have a mash that contains hundreds of flavouring compounds and different alcohols. No different from a Scotch distiller. The difference is that vodka distillers will talk of congeners as "harmful" compounds, which must be eliminated. For whisky, rum and brandy producers, they are the very things that they want to retain to give their spirit its personality.

To get to that stage of cleanliness a vodka will be distilled two, three, four or more times. While some vodkas are produced in pot stills (for example, Smirnoff Black and Ketel One), the majority are made in continuous stills with a rectification column (or columns) to remove the unwanted by-products. The difference between premium vodkas comes in the manner in which they are rectified and, later, filtered.

In the Absolut distillery, for example, the spirit passes through a number of columns, each designed to extract a different set of "impurities". One takes out solvents, another fusel oils, another methanol, while the fourth concentrates the spirit to 96 per cent ABV. Go to the Absolut lab and they'll point proudly to the chemical readouts and show that it's as close to ethanol as you'll get.

Here's the dilemma: distillation and rectification are so efficient that they have also removed the trace elements that give premium vodka its character. Absolut at this stage is indeed absolutely pure – so pure, in fact, that it tastes of nothing. What they have to do is put flavour back in by blending in a spirit that's been distilled at a lower strength, along with some vodka that has been aged in wood. That isn't to say that these top vodkas will blow you away with a massive whack of the original raw material, but it will be there and you can differentiate between them.

Though most vodka producers aim for a neutral rectified spirit, Polish distillers rectify to a lower degree and attempt to retain some elements of the base material, while still achieving purity of flavour and character.

Some distillers are quite happy producing nothing more than ethyl alcohol. These vodkas are stateless drifters that have no connection with the place of their birth – but they have their uses if you are wanting to make a long drink that tastes of the mixer, not the spirit. Other distillers take the Absolut route (though few are willing to share in any aspect of production) to ensure that character and personality are evident in the final product.

FILTRATION

OPPOSITE: *Wood maturation is used for some vodkas*

The final stage for any vodka is filtration, the aim of which is to remove the spirit's raw, aggressive edge and replace it with a mild, mellow, often sweet taste. In many ways, filtration replaces wood ageing as a method of getting smoothness to the spirit.

It's a process that distillers guard jealously – there are as many secret methods of filtration as there are of distillation. The most common method involves passing the spirit through activated charcoal. What the charcoal is made from will have an impact on the spirit – most agree that alder and birch are best – though some distillers use synthetic or bone charcoal, but the results aren't as good. Some have a more complex filtration system. Stolichnaya and Altai, for example, are repeatedly filtered over silver-birch charcoal and pure quartz sand; Suhoi, allegedly, is filtered through diamonds, while Smirnoff is passed through seven columns packed with charcoal. Distillers can then add compounds to round out the mouthfeel.

RIGHT: *The personal touch.*

Technical though this undoubtedly is, at the end of the day the best vodkas are only approved if they are passed by a tasting panel. Any professional spirits taster possesses a rare ability for smell. In vodka they are detecting minuscule differences. "There are people with extraordinarily refined sensitivity," says Dr Skrzypczak. "Before World War Two, there was a Mrs Wasikowa working at the State Spirits Monopoly Central Laboratory. Just by tasting, she could say which of the 50 or so rectification apparatuses then operating in Poland had produced a given sample, and what kind of defect the apparatus had. Another specialist was in the habit of having stiff shots of the tested beverages before starting the tasting proper." You bow in awe.

After filtering, the vodka is either reduced and bottled or passes to another stage – flavouring. Most people these days think that flavoured vodkas are some new phenomenon. Bars have gone flavouring crazy – although they almost inevitably end up with vile home-made examples. Some lower-end brands have also jumped on the bandwagon by adding flavour extracts to the vodka, but these clumsy attempts to replicate the classic old styles are immediately exposed when you taste them side by side. Sadly, though, not everyone has the chance to do this experiment, and these new gimmicky flavours could end up destroying one of vodka's forgotten styles.

While flavours were originally either medicinal herbs or sweetening agents used to disguise off-notes in the distillate, the intricacy of the old Polish and Russian recipes implies that flavour for its own end was an aim at an early stage in vodka's evolution. These classic flavoured vodkas give a window into the past, lighting up the woods and fields that surrounded the early distillers.

Flavouring is added not by redistillation – like gin or akvavit – but by maceration, leaching or, in some cases, by blending in distillates of flavourings or wine. The cheapest option (which also gives the clumsiest examples) is cold compounding, where flavourings are poured into the vodka. To achieve better results you have to macerate the ingredients in the spirit for a lengthy period at room temperature. The time will vary according to the ingredients, and some of the more complex recipes will have different ingredients added at different points.

A newer method, outlined by Faith & Wisniewski in their book *Classic Vodka*, is the circulation process, which is used for styles such as Zubrowka. Here, the flavouring agents are placed on a rack inside the tank and the alcohol is passed through them at regular intervals to get an even, quick extraction. Ricard uses a similar technique for making pastis.

The end result is a vast range of flavours with extraordinary tastes – and a history that dates right back 500 years to vodka's origins. In those days the distillers would have used the ingredients around them – like Tatra vodka flavoured with the herbs from the mountains – while others give a clue to their aristocratic heritage by using the essences and contents of the manor house's spice room.

OPPOSITE: *The new wave of stylish brands*

You'll find vodkas like the Polish winter warmer Krupnik, which is flavoured with honey (the oldest of all fermented drinks) and 30 other herbs and spices, including cinnamon, nutmeg and ginger; or the Russian hunter's vodka, Okhotnichaya, which combines 10 spices and herbs, including ginger, clove, juniper, anise, orange peel and port.

There are echoes of the age-old rural custom of using wild autumn fruits in Jarzebiak, made with rowan berries, or the heady languorous sweetness of wild cherry in Wisniowka. Then there are vodkas which must have originally been the preserve of the rich – lifted, effervescent Cytrynowka (using lemon peel and leaves in Poland) or Russia's equivalent Limmonaya (which only uses peel) and, most expensive of all, the rare, delicate Rose Petal.

You name the flavour, it's there. Bloody Mary aficionados can choose between crisp Pieprzowka; the softer Wyborowa pepper; the dry Absolut Peppar; or the powerful Stolichnaya Pertsovka. To put a different spin on things, there is even oak-aged vodka, Starka. In Poland this is made from a 50 per cent ABV rye spirit that is aged in Tokaji wine barrels or large vats, sometimes with a touch of Malaga fortified wine to sweeten it. Originally, this was a feast vodka, which was made by pouring the spirit over the lees in a wine barrel and then burying it for three to

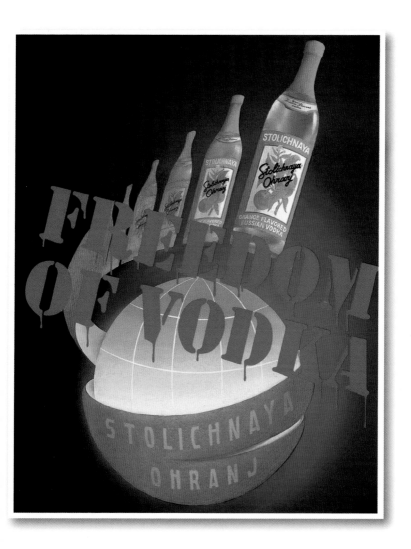

ABOVE: *Part of the new wave of post-Communist vodkas*

four years. Starka is pretty rare, but those keen to explore them should investigate the range from Szczecin distillery – Specjalna (12-year-old), Jubileuszowa (15-year-old), Piastowska (20-year-old) and Banquet (30-year-old).

Russian Starka has apple and pear leaves, as well as a fortification of brandy and port. Unusual though it may seem to those of us brought up on naked vodka, adding wine and other distillates is common practice. In Poland the marvellous Zytnia is a strong rye vodka which has had apple and plum wine blended in.

If that leaves your mind reeling, unsure of where to start, then go straight to the top of the tree, and search out Zubrowka (spelt with a 'v' in Russian).

Hailing from the Bialowieska forest on the Polish/Belarus border, Zubrowka was originally a seventeenth-century regional speciality, and a particular favourite of the Polish royal family on their hunting visits to the forest. Each bottle of Zubrowka has a blade of bison grass (hierochloe odorata) in it. This plant is the favourite grazing of the European bison, which still roam wild in the forest – in fact, legend has it that only grass that's been urinated on by one of the beasts can be used in the vodka. The grass itself is high in fragrant components which impart an evocative, green scent to the vodka with a delicate vanilla touch (coming from the coumarin ester in the grass). A glorious drink.

OPPOSITE: *Lemon appeal*

POLISH VODKA

"No, bo wodka stygnie!" (Hey, the vodka's getting cold!)

Polish toast

ABOVE: *Vodka started life as an apothecaries' potion*

It's impossible to wrap up vodka's history in one neat bundle because it has evolved differently in each of the countries where it is the national spirit. Vodka may be regarded as a commodity by the West, but in Poland, Russia and Scandinavia it is a very different beast.

Trying to get to the bottom of Polish vodka alone is a Herculean task – with an estimated 1,000 brands available, it's a mission that would take a lifetime. Poland is the best place to start when looking at vodka's history. It may not please Russians or Swedes, but there is convincing evidence that the secret of distilling (wine initially) filtered into Poland from the West and spread from there into Russia and the Baltic States.

Peasants in the eighth century were making a crude alcoholic spirit by freezing wine, though the first written record of a spirit made from grain comes in 1405 – predating Russia. Precisely how distillation arrived in Poland is a matter of conjecture, but it is likely that the secret was brought to Poland by Italian monks (although some suggest they could have been Irish) who were, by then, well versed in the arcane subject of transforming a base material into another, far more potent, one.

As in every other country, these early spirits were initially used as medicines. In 1534, Stefan Falimirz devoted a chapter in his herbal book to distilling vodkas, but their usage was limited either to cures or: "for cleansing the chin after shaving [or] rubbed on after washing in the bath". Even in its earliest incarnation, vodka was being used as something rather stylish – an aftershave or cologne. Falimirz, however, claimed that not only did vodka make people smell nicer, but it also could be used "to increase fertility and awaken lust". Right enough, perfumes these days still allude to that latter quality.

These early vodkas wouldn't have been the clear, neat spirit we know today. They would have been spiced up with infusions of herbs, roots, spices and exotica like marzipan, almonds and sugar. The result was that, according to Marian Hanik's *History of Vodka in Poland*, the Polish pharmacies of their time were more like cafes, where people came to take their medicine and have a chat.

Vodka was given a significant boost when, in 1546, King Jan Olbracht decreed that all Poles could make alcohol – which they gleefully did, even though they had to pay tax for the privilege. Although this decree was soon restricted to the gentry, they seized on the money-making opportunity.

By the end of the sixteenth century, Poland was producing vodka in sufficient commercial quantities to start exporting – a trade that would grow in importance over the following two centuries. Production initially centred round the then capital of Cracow, and by 1580 there were 500 distilleries in Poznan, while Gdansk had taken over as vodka's capital – its first distillery being started by a Dutchman called Ambrosius Vermoellen.

With production rights restricted to the upper class, distilleries were springing up across the country, in towns, monasteries and in country houses and estates. "Naked" vodka was still less common than the vast range of flavoured vodkas that had appeared on the back of this distilling fever – there were medicinal vodkas, country vodkas flavoured with wild herbs, sweet vodkas (like Krupnik) for winter. Faith & Wisniewski, in their *Classic Vodka*, recount that over 100 different flavoured varieties were being produced in this period. Vodka had become part and parcel of Poland's life. Everyone drank it and, despite the legal restrictions, everyone made it, though quality differed according to your wealth. Those who could afford it produced vodka from rye, while the peasants had to make do with anything else they could get their hands on.

Hanik gives a typical example of a Polish manor house distillery in the seventeenth century – it drew water from local springs, had a malt house, a mill, cooperage and smithy; an ice house for cooling the yeast and the mash; a still room and a barrel warehouse. It was a sophisticated operation, not some cobbled-together, moonshining unit. The spring water was filtered through charcoal, distillation was in copper stills and redistillation was common (though not universal).

With the arrival of triple (and, on occasion, quadruple) distillation and charcoal filtering in the eighteenth century, came a new style of vodka that was stronger and cleaner. Polish vodka became the model for quality production across Eastern Europe with equipment (and techniques) being exported to Russia and Sweden.

Vodkas would have been made from the main Polish starch crops – rye, wheat, barley and oats were all used. Potatoes, although they had arrived as an exotic ingredient in the fifteenth century, only began to be used for vodka production in the middle of the eighteenth century, becoming a major raw material a hundred years later.

The style was still following the aim of the earliest distillers, to try to produce as clean and pure a spirit as possible. Vodka by now was not only a spirit that could be drunk on its own, but one that would also provide a base for the seemingly infinite number of flavourings that had become established as an integral part of Polish vodka's style (see below).

The next major step towards this goal of a pure spirit came at the start of the nineteenth century, when the first three-chambered Pistorius column still was installed at General Ludowik Pac's distillery. When steam was incorporated into the method in 1826, Polish vodka production was ready

OPPOSITE: *The continuous still improved quality immeasurably*

to expand once more – with higher volumes of purer spirit flooding on to the market.

Although rectifying columns didn't make their first appearance until 1871, Pistorius equipment was a major breakthrough. From here on in purity was not only the aim, it was an achievable goal. As well as new technology appearing, as Faith & Wisniewski point out, the emergence of the first kosher vodkas in the 1830s also had a significant impact on improving distilling practice and making the production process as clean as possible.

The combination of this new technology and the planting of potatoes as a main crop meant that the beginning of the nineteenth century saw a doubling of vodka consumption and an explosion in the number of distilleries. It wasn't to last long.

The potato blight (1843–51) crippled spirits production as it did in Ireland – while increased taxation put paid to most of the smaller distilleries. By the end of the century, rural distillers could no longer compete in terms of volume or quality. Although they produced spirit, they were sending it off to the large plants for rectification. This spirit was either bottled or

LEFT: *Vodka has always played a major part in Polish social life*

sold on once more to producers specializing in flavoured styles. These producers were using the new Henckmann equipment, which used alcohol vapour to strip flavours and speed up the process.

During the First World War, there was a further fall in the number of legal distilleries – much of it caused by a further rise in taxation, fuelling a rise in home distilling. In the 1930s, 4,000 illegal stills were confiscated every year. Vodka was too important to Polish life and culture to be given up without a fight. The industry, however, consolidated further and was brought under the control of a state monopoly after the Second World War. In 1973, this body became Polmos.

Vodka remained the lifeblood of Polish society. During the shortages of the 1980s it was used as a form of currency to barter for goods, and those who wanted more than their ration made their own. Wisniewski highlights the sudden increase in the use of junior chemistry sets during this desperate time. With the arrival of democracy, the 25 distilleries controlled by Polmos were granted independent status, although they are still ostensibly government controlled.

This has produced an explosion of brands – there are an estimated 1,000 brands of Polish vodka available – with each distillery trying to find its own point of difference. It's only to the benefit of the spirits lover, as the newly liberated distilleries have gone for the top end of the market.

OPPOSITE: *A Polish bestseller.*

It's difficult to describe how important a role vodka plays in Polish and Russian culture. It's not just a neutral spirit to be diluted with a soft drink; it's a social event with its own rituals. Flavours aren't the recent invention that the West seems to think they are – they all have their own use, their own ethos.

In Poland, drinking has long been seen as a social event, an act of generosity and hospitality. This means it's regarded in a very different light from how it is in the West (and, in particular, the USA). Hanik quotes Jedrzej Kitowicz, who wrote in 1850: "Among the Poles, nothing could be done without getting drunk... It was the host's greatest ... satisfaction when [the day after a party] he heard how none of his guests had left sober."

It was nothing new. "Among them [the Poles] getting drunk is a praiseworthy custom, certain proof of sincerity and good manners," wrote Fulvius Ruggeri to Pope Pius V in his 1568 description of distilling in Poland. The maxim was repeated 400 years later by the aristocrat Czartoryski, who wrote: "twice a year, one should get properly drunk".

It might seem reprehensible advice in these abstemious times, when the merest hint of a lack of sobriety is seen as a crime, but vodka remains a safety valve in Polish life, as well as an inherent part of the culture.

Poles, Russians and Swedes all treat vodka in the same way as the French treat wine. It's an aperitif, served with the snacks that appear before the meal, drunk as a liqueur after the meal. Children are weaned on to its charms at an early age – and end up more responsible drinkers as a result. My (half-Polish) brother-in-law remembers as a child being given a small glass of Wisniowka (a sweet, cherry vodka) on special occasions.

Polish vodka's inherent high quality and versatility demand to be examined more closely.

STOLICHNAYA
Cristall

From the Cristall Distillery,
Moscow, comes the world's
finest Russian Vodka,
Stolichnaya Cristall,
ultimate in smoothness,
fine grain neutral spirit,
distilled and bottled
in Russia.

RUSSIAN VODKA

———◆·✕·◆———

"The first glass of vodka goes down like a post, the second like a falcon and the third like a little bird."

Russian proverb

It's impossible to wrap up vodka's history in one neat bundle because it has evolved differently in each of the countries where it is the national spirit. Vodka may be regarded as a commodity by the West, but in Poland, Russia and Scandinavia it is a very different beast.

If Poland can lay claim to being the first vodka producer, its image is still associated with Russia. But until the mid-fifteenth century the Russian nobility were still sipping on mead and wines, while the people were drinking beer. Faith & Wisniewski argue that some distillation was taking place beforehand as an adjunct to producing pitch from pine. It's possible that alcohol was being distilled in log stills (the same thing happened in Kentucky), and it was the need to keep re-distilling the alcohol to remove unwanted elements that, they argue, established vodka distillers' obsession with producing a pure spirit.

This proposition goes a long way to explaining why, when other distillers across Europe were happy to retain flavouring compounds, those in Russia and Poland wanted to get rid of as many as possible. There is a further theory, but we'll come to that in a minute.

Russian vodka historian William Pokhelbin claims that vodka production was known from the mid-fifteenth century, citing the sudden degeneration of public morals and accounts of mass drunkenness and violence. This, he argues, points to a change in the type of alcohol consumed. Sadly, whether this wild depravity was caused by strong spirits or another wider social cause we cannot (yet) ascertain.

Though there is no evidence that vodka was at the root of this dissolute behaviour, records do show that a Russian delegation had visited Italian monasteries in 1430 where they were shown how to make aqua vitae. Given that their return co-incided with a grain surplus, it is entirely possible that vodka distillation was in place by the middle of the century.

Virtually every country has tried to control spirits in some way or other. Taxation is the most common method. In Russia they went one step further. While Jan Olbrecht was allowing all Poles the right to distil, Ivan the Great had already established the world's first spirit monopoly. Ivan the Terrible took state control one stage further, decreeing that vodka could only be

sold in official taverns and be produced from stills owned either by tavern owners or nobles. Vodka fitted in neatly with the age-old framework of Russian society – the landowners got rich and the poor got drunk.

ABOVE: *Queuing for the vodka ration*

Though Peter the Great (1672–1725) liberalized distillation – mainly to collect taxes – vodka production was a rich man's hobby. Peter himself invented a modified still to improve quality, and his recipe for vodka involved producing a triple-distilled spirit, flavouring it with anise and then redistilling it once more. The search for a pristine spirit was well advanced.

It could be argued that this was only to the benefit of vodka's quality. By the turn of the seventeenth century, four distillations were common, and exotic, complex flavoured vodkas were being drunk by the ruling classes. Filtering the spirit through charcoal to further clean it up arrived in the eighteenth century – at much the same time as in Poland – and is credited to Theodore Lowitz, who was working on the Tsar's request. There's every chance that it was a Russian or Polish émigré who took the art of charcoal filtration to America and into Tennessee.

It certainly was known long before the 1880s, when Smirnoff claims to have invented the technique.

In the eighteenth century, a series of government bills restricted production still further. This effectively split production into two quality tiers. The nobility, who used their vast feudal power and wealth to make it to perfection; and the state distillers, who supplied base spirit for the poor. While the poor soaked up crude spirits to obliterate their misery, the top Russian vodkas began to acquire an international reputation. Catherine II sent vodka to the kings of Prussia and Sweden, to Voltaire and to her friend, the Swedish botanist Karl von Linne,

RIGHT: *Catherine the Great: promoter of vodka*

who, suitably inspired, wrote a lengthy treatise on "vodka in the hands of a philosopher, physician and commoner".

By now, class divisions had created different vodkas for each stratum of society. The nobility drank rye and elegantly flavoured vodkas, while the poor drank rank spirit made from potato, beet and nettles bought in at low prices from Poland, Germany and from illicit stills, rather than from the state distillers.

The state was forced to take control of production once more, cut the number of distilleries in half, introduced the continuous still and, under the controlling influence of the chemist Dr Mendeleyev, set quality control standards. By the time that Piotr Smirnov started to make his own vodka in 1861, the best Russian vodkas had become the elegant, clean spirit we know today.

Smirnov's rise was inexorable. The firm built a distillery in Moscow in 1868 and was granted a Royal Warrant in 1896 – an award which didn't go down too well with the Bolsheviks. By the turn of the century, the firm was producing 3,500,000 cases a year and had an annual income of US$2 million.

Vodka was very big business and distillers were pillars of the capitalist class. Little surprise that vodka was one of the first targets of the Bolshevik government – and that Vladimir Smirnov fled to Constantinople, Lvov and Paris in his unsuccessful attempt to start all over again.

The initial reaction of the Bolshevik government was to restrict vodka's strength to 20 per cent ABV to try to reduce drunkenness – but people just kept on making stronger stuff at home.

Things changed dramatically when Stalin took power. Faith & Wisniewski make a strong case that Stalin used vodka as a means of social control. By keeping the price artificially low and making it easily available, he ensured that the Soviet Union was in a state of endemic alcoholism. Vodka was a way of suppressing dissent and, in those dark days, the drinking wasn't heroic – it was desperate.

Stalin wasn't the first to use cheap alcohol as means of keeping people compliant. Rum was also used in this fash-

ion in the Caribbean by planters who gave the spirit to their slaves. Brandy, too, was given to the black majority by white employers in apartheid South Africa.

Now with Communism (allegedly) overthrown, things have gone full circle, with Smirnoff now being produced in pot stills in Moscow once more. The market has opened up, allowing Stolichnaya to blast on to the world market. Then there are Moskovskaya, Kubanskaya and Ultraa from St Petersburg, and vodkas like Altai, a winter wheat vodka from Zmeinogorsk in Western Siberia. There may not be quite as many brands as there are in Poland, but the gates have been opened.

Vodka epitomises Russia. It has kept spirits buoyant in times of desperation and obliterated misery. In the wilds of the Altai and Tuva, it is used by shamans as a libation to the spirits. In Moscow it is sipped by the new bourgeoisie.

Although Russia's future is less than clear, one thing is for certain: empires may rise and fall, but vodka will always survive.

LEFT: *The cream of the Russian crop*

SCANDINAVIAN
VODKA

———◆◇◆———

"No matter how much you drink, there will always
be booze in the world ..."

Finnish proverb

Records suggest that distillation arrived in Sweden in the fourteenth century when brannvin (burnt wine) was being distilled from imported wine – and domestic grain.

This was initially used as a medicine (which has echoes in later years) and also in the development of gunpowder. It was first embraced by the aristocracy, and only began to spread to the rest of the population in the seventeenth century.

As soon as it did, though, distilling became endemic. By 1756, the country could boast 180,000 stills.

Home distillation was banned in 1860, the first of many attempts by the Swedish government to exert control over consumption, and large commercial plants equipped with Coffey stills began to exert control over the market – and produce more vodka. Less than 20 years later, Sweden's most famous distiller emerged on the scene.

Lars Olsson Smith started distilling young. Unbelievably, by the time he had reached his early teens he was producing an estimated one third of Sweden's vodka. It was his creation of Sweden's first rectified spirit in 1879 that was to elevate his name to one of legend.

Smith was a stubborn individual with very fixed ideas about quality. He was so convinced that his new Absolut Rent Brannvin (Absolutely Pure Vodka) was the very best vodka on the market that he took on the might of the Stockholm monopoly.

His distillery was located in an elegant house situated on the small island of Reimersholme, conveniently just outside Stockholm's city boundaries. By providing a free ferry service to his customers he not only guaranteed high sales, but angered the other distillers to such an extent that shots were fired at the boats.

His Absolut vodka was an instant success and Smith soon needed more raw material for his new brand. Accordingly, he switched production to the far south of the country, to the wheat fields of Skåne. He went about buying up distilleries and ensured that his vodka was the best-distributed brand in the country – at one point even using the unions to boycott shops that were selling what he claimed were inferior brands.

By the end of the First World War, Sweden had a state monopoly, Vin & Spirit, in place to control the sales and production of all alcoholic beverages and immediately put up taxes to astronomical levels.

Who knows what would have happened to the fortunes of Smith's brand had it not fallen victim to Sweden's strict state control of drink?

Scandinavian governments (with the notable exception of Denmark) have long had a strange relationship with alcohol. The people drink a lot of it, but their rulers don't like them indulging and, until recently, have tried everything in their powers (Finland even tried a period of total prohibition) to make it impossible to enjoy a drink – though without any real success.

It only encouraged Swedes, Finns and Norwegians (who also have state monopolies) to go on massive drinking binges, either on the duty-free ferry routes between the countries, or in Denmark. Many people have also continued to distil at home. Even the Swedish state monopoly will admit that more than 20 per cent of the vodka drunk in Sweden is distilled privately – though the true figure in fact could be a great deal higher.

BELOW: *Spirit of the mountains*

Strangely, this tight control – which is part and parcel of the Swedish state's benign interpretation of socialism – created one of the world's biggest spirits brands. Absolut was forgotten until 1979 when V&S decided to hit the international vodka market.

The brand was distilled once more, repackaged in a replica of an old medicine bottle and went for the top end of the US market – the most competitive vodka market in the world. Carillon Importers' Michel Roux took it up and ran with it (he was later to do the same with Bombay Sapphire). It's now the seventh biggest brand in the world.

Absolut was following in the footsteps of Finlandia, which had hit the USA in 1970. Finnish vodka dates back to the sixteenth century, but it too was taken under the state's wing in the twentieth century – when the distillery was closed down to make alcohol for Molotov cocktails.

Denmark traditionally has been better known for its akvavits, but has always produced fine quality vodkas. The father of these was Isidor Henius who, in the 1850s, installed the country's first rectification column at his Aalborg distillery.

This famous site is now owned by the giant producer Danisco, home to the new wave of Danish vodkas foremost among which is Danzka. To be strictly accurate, Danzka was originally produced by an independent distiller but, in the way of these things, was snapped up by the giant firm when it saw the potential for a high-quality brand on the export markets. The fine, ultra-clean grain vodka in its distinctively packaged silver metal bottles comes in a range of flavours, including Citron and Currant. They are fine and rather attractive. The firm also produces Fris, a vodka created to be served straight from the freezer.

Norway's contribution to the growing band of Scandinavian brands is the potato vodka Vikingfjord. It's a good example of the Nordic style, flirting with neutrality, but being saved by smooth mouthfeel and delicate flavour.

Finnish industry started in earnest in the 1950s and now

has one of the most technically advanced distilleries in the world making the Finlandia a by-word for Scandinavian purity.

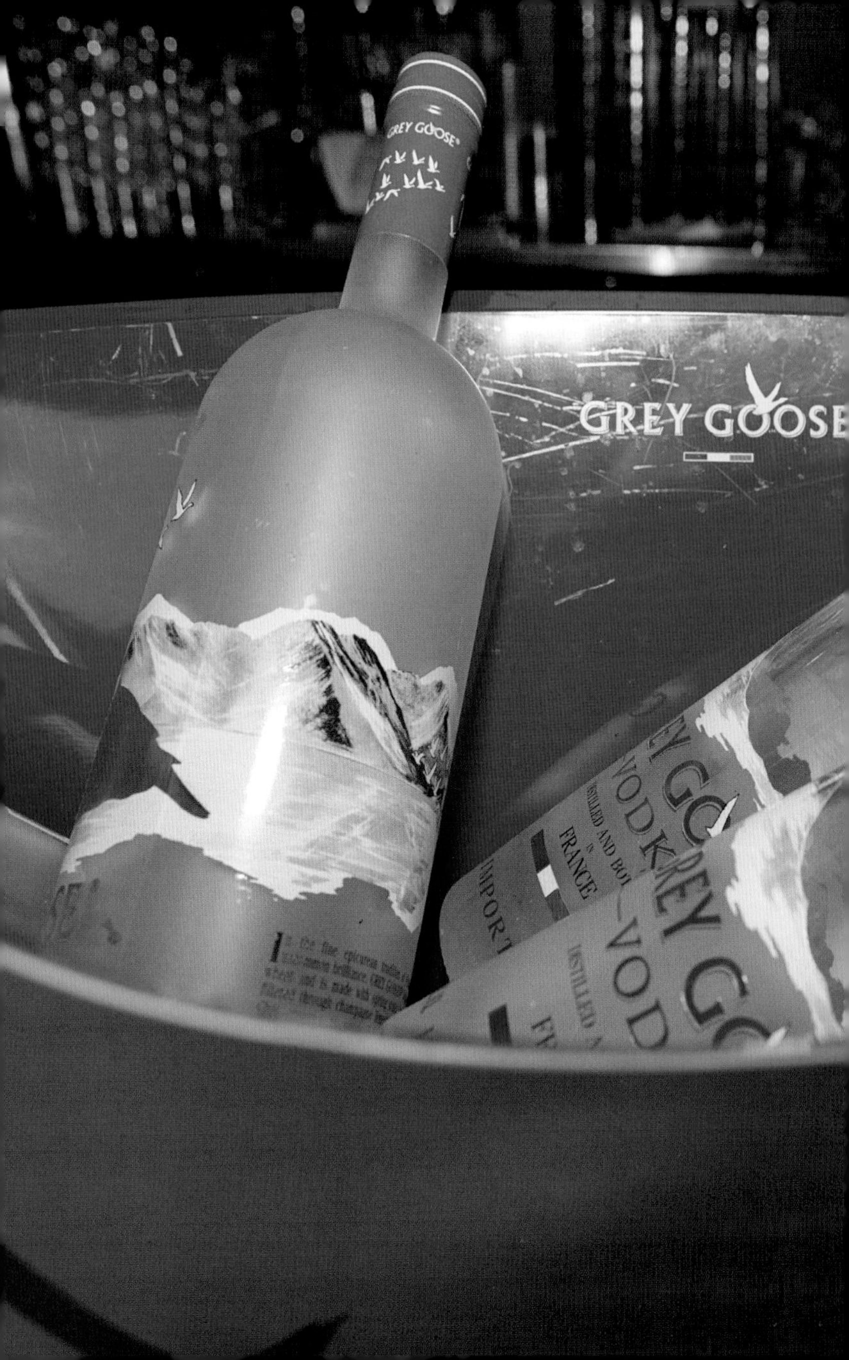
GREY GOOSE®
GREY GOOSE
FRANCE

VODKA
FROM THE
REST OF
THE WORLD

"The problem with the world is that everyone
is a few drinks behind."

Humphrey Bogart

The rest of the world has always looked on vodka in a slightly different light to eastern Europe and Scandinavia. Although you can say that vodka akvavit, korn and gin are all white spirits from the northern European grain belt, they have evolved in subtly different ways.

In vodka, the prime motivator was to produce as clean a spirit as possible. This could then either be drunk neat or subtly flavoured. While cleanliness of the base spirit was important for the rest of these white spirits, their aim was to still have a dominant powerful character – juniper for gin, caraway for akvavit.

RIGHT: *The sky's the limit. Part of the Campari stable, Skyy Vodka now comes in a range of flavours.*

It wasn't until America recreated vodka as a neutral, mixable spirit in the 1950s that vodka began to make inroads into Western European drinking culture. Vodka had been produced prior to this date – the Dutch firm Hooghoudt has been making vodka since the end of the last century – but it had never captured the public imagination. Now, with America reinforcing its role as the arbiter of global cultural taste, vodka became the latest in a chain of cultural signs like jazz, rock n' roll and the movies, which signified modernity – and "American-ness".

The vodka that Western Europe drank therefore wasn't a European creation but an American one, a drink where neutrality was the sole intent. But what's the difference between any of these commodity vodkas with their pseudo-Tsarist names and Moskovskaya, Luksusowa or Belvedere? The simple answer is that the quality brands have retained subtle traces of their raw material, they have grace and elegance – and they therefore retain your interest.

What you got with the Westernized versions was a spirit that, even neat, gave no offence, that was a bland, non-commital mixer. Vodka was dumbed down before the phrase was ever thought up.

It was also relatively easy to produce. Any distillery with a rectifying column can make it. Smirnoff is made across the globe, there's Suhoi from Italy, Zar from Bolivia. Things, however, are beginning to change and a new, more flavoursome premium sector is emerging. There's Skyy and welcome innovations such as Tito's Texas Handmade and the organic Rain from the Ancient Age distillery in Kentucky, Fris and Danska from Denmark and the Dutch brands Ketel One and Royalty. The West is slowly beginning to pay attention and take note of what real vodka is all about.

RIGHT: *Medal winner: Tito's Handmade Vodka, which is made from yellow corn, has won plaudits from both the industry and critics.*

BRAND DIRECTORY

TODAY'S MARKET

Vodka will endure, of that there's little doubt. The question is whether it can break free of its commodity image and be recognized as the classy spirit it is. There are a number of different factors at play in today's market. Almost inevitably, there has been a raft of brands that have entered the market hanging on to Absolut's coat-tails. Here style is considerably more important than content. Fancy bottles and huge price tags can't hide the fact that there are a large number of premium vodkas on the market that are not worth the money, and hopefully someone is soon going to notice that these Emperors (or Tsars) have no clothes.

At the same time, though, there are "new" brands from Poland and Russia that deserve wider attention. Vodka lovers already appreciate the herbal lift of Stolichnaya (and its richer big brother, Cristal), Absolut's clarity and Finlandia's light dryness.

But don't pass over the lime oil richness of Wyborowa; Moskovskaya's full, elegant, rye weight; and the creamy, soft Luksusowa – or dismiss Smirnoff Black. Poland can also weigh in with the premium potato vodkas Baltic and Chopin, or the rye crunch of Belvedere. Also try and find the kosher vodkas made at the Nissebaum plant in Bielsko-Biala or at Lancut – it's worth it.

Elsewhere you can browse among Holland's pot-still Ketel One, or Royalty, Denmark's Fris and Danska and, from the States, Rain and Tito's Texas Handmade. Though these may all be seen as esoteric brands at the moment, they are providing the impetus for vodka's future long-term success as a serious spirit.

This shift upmarket is much needed, as the category has become increasingly bogged down with inferior, cheap products and a rather peculiar image. Vodka isn't seen as a bad spirit, but few consumers in recent years have bothered giving it a second thought. This in itself isn't surprising as, since the 1950s, the whole category has been sold on the premise that it tastes of nothing.

OPPOSITE: *Fris celebrates a Viking heritage.*

Well, as we have seen, times are changing and vodka is most definitely making a comeback.

POLISH BRANDS

RYE-BASED

Belvedere Dry and clean with a dusty rye edge that's mixed with dried lime and coriander. Cool and fresh. The palate mixes caraway and clean acidity.

Bols Lightly mineral nose. A pure, clean and soft palate that offsets a delicate sweetness with a hint of sootiness on the end.

Sobieski A top-seller in its home country, this rye vodka has a gentle nose with hints of rye tartness and fruits. Very clean, with a soft sweetness in the centre balanced by a sour, peppery rye finish.

TASTING NOTES

BRAND	AROMA	TASTE	RATING 1-5

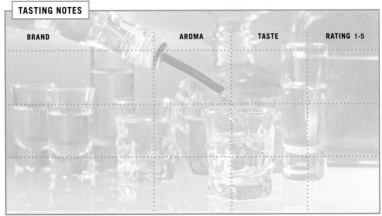

Wyborowa Blue A classic old rye vodka which was the first Polish vodka to gain credibility in the West. With a mix of almond and pepper on the nose, the palate manages to mix an oily mouthfeel with zingy spices and a light touch of allspice.

Wyborowa Exquisite The firm's super-premium expression has a softer nose, with the sweet/sour rye bread accents being more restrained. The palate is considerably more crisp and dry, with greater nutty crunch and higher levels of spice and an almost salty edge.

POTATO-BASED

Luksusowa Red Label Potato-based vodkas are another Polish speciality. A very gentle nose with a slight vegetal character gives way to the creamy/buttery palate typical of potato vodkas. Sweet with some fennel on the finish.

Chopin Gentle and discreet nose. While this has the umami-rich palate of potato vodkas, it is balanced with a clean peppery spice and a cereal-like crunch.

TASTING NOTES			
BRAND	**AROMA**	**TASTE**	**RATING 1-5**

MIXED BASE

U'luvka Made from a mix of wheat, barley and rye, this has a lifted, quite aromatic nose with distinct graininess. The palate is almost perfumed, with a lime-like zest, a succulent mid-palate and a driving spicy finish.

Soplica A mix of wheat and rye is used here to produce a vodka which has a forward, aromatic nose with cumin-like spices. This discreet fragrant character continues in the mouth, where a soft mid-palate gives way to a zingy, energetic rye-accented finish.

U'LUVKA
VODKA

TASTING NOTES

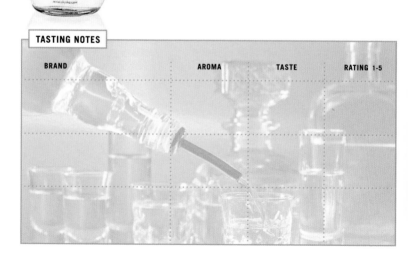

BRAND	AROMA	TASTE	RATING 1-5

RUSSIAN BRANDS

Beluga Barley-based Siberian vodka with a firm, slightly nutty nose and hints of basil leaf. The palate is rich and smooth with a firm grain undertow and a lengthy clean finish.

Cristall (Russia) Pure, ultra-clean grainy nose. Elegant, delicate and one for sipping.

Etalon Hailing from Belorussia, this rye/wheat vodka is crisp and very clean with a light and scented nose akin to green apple and tree sap. The palate is soft and silky with a fragrant aniseed/liquorice note.

TASTING NOTES

BRAND	AROMA	TASTE	RATING 1-5

Green Mark Another 100 per cent wheat vodka. Once again the Russian oiliness is apparent on the nose alongside a light hint of fennel. The palate has some weight. Green olive in the middle, then a bone-dry finish.

Ikon Fresh and clean on the nose and, like Etalon, with a little pine edge. A focused and quite intense needle on the palate adds interest, while the finish brings out a touch of caraway and pepper.

Imperia The super-premium expression of Russian Standard is wheat-based and has a very clean, fresh nose. The palate is rich and textured with white pepper, anise and a hint of citrus. The finish is medium-length and clean.

Kauffman Luxury Vintage Unusual insofar as it is only made in years in which the distiller deems the wheat to be of suitable quality. Here you meet a clean, gentle nose with undercurrents of allspice and anise. The palate is ever so slightly herbal with a menthol coolness, leading to a long, clean finish.

Stolichnaya Elit The super-premium variant of the old classic is wheat-based and is triple-filtered. The nose is restrained and cool with a mineral quality. In the mouth, however, it has weight and depth, with a little sweet spot in the middle before the tight, fennel-like finish.

Zyr Produced from a mix of wheat and rye, here is a vodka which is lightly sweet on the nose with a touch of cereal. The mouth has the oiliness typical of Russian brands but with a firmness given by wheat underneath leading to a steely finish.

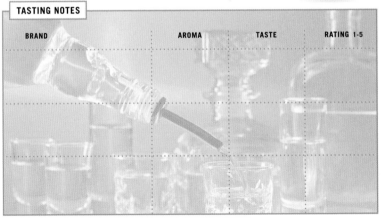

TASTING NOTES

BRAND	AROMA	TASTE	RATING 1-5

TASTING NOTES

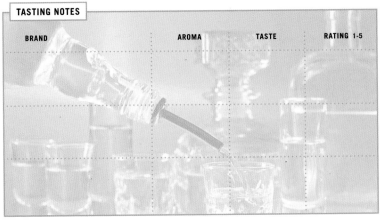

BRAND	AROMA	TASTE	RATING 1-5

SCANDINAVIAN BRANDS

Absolut (Sweden) Clean, light and neutral. Although a style icon, the arrival of other higher quality vodkas has rather exposed its lack of complexity and character.

TASTING NOTES			
BRAND	AROMA	TASTE	RATING 1-5

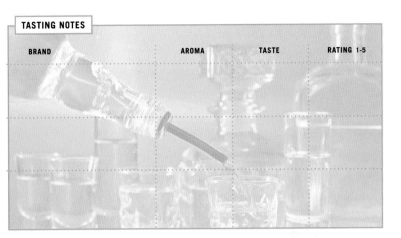

BRAND	AROMA	TASTE	RATING 1-5

Finlandia (Finland) An excellent wheat vodka, clean with hint of lime and citrus and a fresh bite on the finish.

Danzka (Denmark) Strange bottle, which may slightly cheapen what is actually a very decent vodka. Made by the same firm behind Aalborg akvavit (and Cherry Heering), it is a whistle-clean, pure spirit. It doesn't have quite the character and complexity of many from Poland and Russia but it is a high-quality everyday brand.

Fris (Denmark) Good weight, although it seems glycerol-rich in texture. Like all these vodkas, the nose doesn't reveal much; this is all about palate weight.

Vikingfjord (Norway) A rarity, a Scandinavian potato vodka. Typically light and clean on the nose but good texture on the palate.

TASTING NOTES

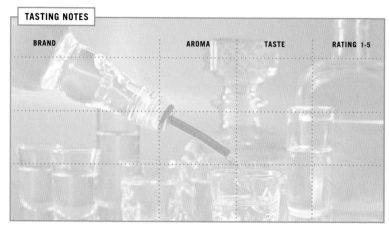

BRAND		AROMA	TASTE	RATING 1-5

OTHER BRANDS

Chase An English vodka made from potatoes, this has the typical buttery creaminess on nose and tongue that you would expect, but with a hint of fruit behind and a clean, spicy finish.

Ciroc French-made, grape-based, this is one of the most aromatically lifted of the new wave of vodkas with touches of vetiver, lemon and nettles. Sweet and almost fizzy spiciness on the tongue and a long citric finish.

Crystal Head The packaging (the clue is in the name) is distinctive, as is the acetone-heavy aroma with touches of bubblegum in the background. The palate is clean and quite firm.

42 Below New Zealand vodka made from wheat and with the anise/liquorice nose typical of that grain. The palate is fine with decent depth and a lightly peppery finish.

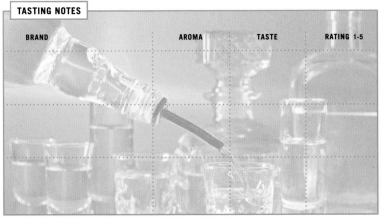

TASTING NOTES			
BRAND	AROMA	TASTE	RATING 1-5

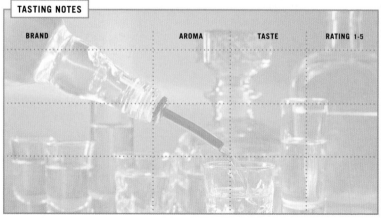

TASTING NOTES

BRAND	AROMA	TASTE	RATING 1-5

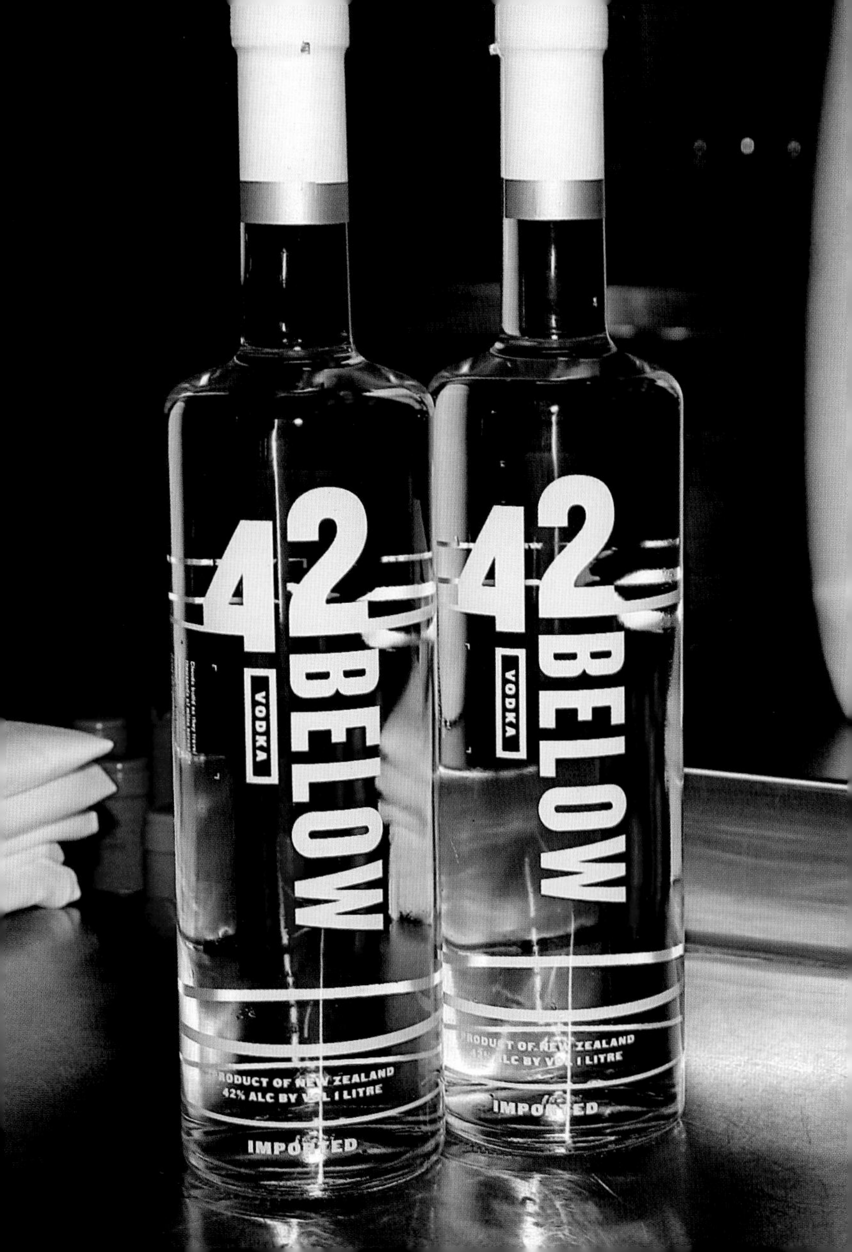
42 BELOW
VODKA
42 BELOW
VODKA

and pot stills; this is one of the fuller-bodied vodkas on the market. Lightly floral with a clean minerality and lime zest. The palate has dry cereal underpinning a silky mouthfeel. The crisp finish palate hints at charcoal.

Rain (US) Organic vodka from Kentucky. Soft and sweet with a hint of liniment and spice on the nose. Chewy, but beautifully clean and wheat-soft with apple on the finish.

TASTING NOTES

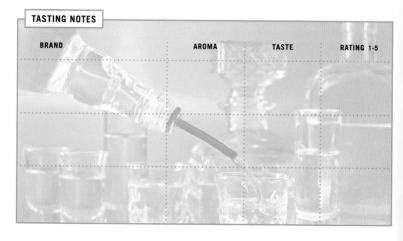

BRAND	AROMA	TASTE	RATING 1-5

Reyka Distilled in Iceland from wheat, this is a very soft and gentle vodka with a fleshy nose. The palate picks out subtle pear sweetness alongside liquorice and a tingling spiciness on the finish.

Smirnoff The biggest-selling "Western" vodka in the world, Smirnoff is distilled at various locations globally. It also often outperforms more pricey competitors at blind tastings. It is grainy on the nose, clean and plump on the palate.

Van Gogh Blue Distilled in Holland from wheat, this has quite a weighty nose with little hints of the sweetness of a grain loft. A pastoral palate with good thickness in the middle of the tongue, it finishes with typical wheat-derived focus.

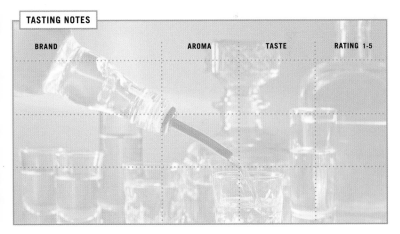

TASTING NOTES			
BRAND	**AROMA**	**TASTE**	**RATING 1-5**

FLAVOURED VODKA: POLISH

Korzen Zycia A ginseng vodka with a peculiarly green/stewed nose.

Krupnik (honey and spice) Very old style of flavoured vodka. Rich, sweet and herbal. Try it warmed up.

Wisent Bison Grass (rye) One of the best – fragrant, elegant and gorgeously perfumed. It is free of the natural flavouring coumarin, to which the US government objected.

Wisniowka (cherry) Concentrated cherry notes, off-dry, with a lovely kick halfway through.

Zubrowka (bison grass) Mown grass, lavender, flowers and spice. Complex, long and sophisticated. The greatest flavoured vodka style of all.

Extra Zytnia (rye vodka, apple spirit and fruits) Clean and spicy thanks to the rye with a perfumed delicate palate. Lovely balance.

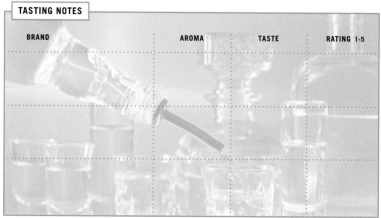

TASTING NOTES

BRAND	AROMA	TASTE	RATING 1-5

FLAVOURED VODKA: DUTCH

Ketel Citroen Natural and juicy with some dusky floral notes behind. Oily texture with delicate lemon all the way through.

Vincent Van Gogh Citroen Highly citric and zesty. Use carefully to give cocktails an extra layer of complexity.

Ursus Citrus Some nice citric notes, but lacks depth.

Ursus Roter A sloe vodka. Dark, but lacking the sweet, bitter interplay which would give it complexity.

TASTING NOTES			
BRAND	AROMA	TASTE	RATING 1-5

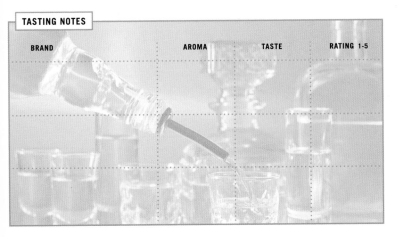

VODKA COCKTAILS

"I like to have a Martini,
Two at the very most.
After three I'm under the table,
after four I'm under my host."

Dorothy Parker

VODKA COCKTAILS

Isn't it interesting how the spirits are raised whenever the invitation slides through the mail slot? "Come over for cocktails. 7 p.m. Saturday." It must be the expectation of the taste of exotic things to come that causes the heart to flutter and the lips to pucker up.

Give 10 people a large bottle of vodka, a selection of mixers, a cocktail shaker and bags of ice and it is guaranteed that within minutes someone will claim they can make the best Daiquiri this side of Cuba – and would you like to try one?

This section is for those who love the cocktail, and to allow them to make and serve a cocktail in the right manner. Not for them the bottles or sachets of premixed horrors sold as Margaritas or Moscow Mules. A handmade cocktail, stirred, mixed, or shaken, is a work of art descended from a great tradition.

These recipes are the real thing, a selection of classic and modern cocktails as they were meant to be mixed and served. There are also modifications to recipes acknowledged as the original; sometimes the additional ingredients are an improvement, created by a bartender who wasn't satisfied with the taste of the original. It is only through experimentation that we can seek perfection.

Whether you are a whisky lover, or a vodka-on-the-rocks fanatic, you will be pleasantly surprised at the selection of recipes here. There are sweet, sour, creamy, frozen, purely spirit, and bubbly recipes in this section, served in a selection of glasses: highball, tumbler, old-fashioned, Margarita, champagne flute, liqueur and shot glasses. However, it is the elegant martini (cocktail) glass that best serves most recipes. It's the taste and colour of what lies inside its sleek shape that matters.

Without vodka there would be no Cosmopolitan, Apple Martini, or Bloody Mary. Its first appearance in America came in the 1950s when it was mixed with ginger beer, thus creating the Moscow Mule, which was served in a copper mug. Vodka is a great mixer, doesn't colour the drink, or taste strong on the breath. New vodkas are not as tasteless as they once were. Here is a selection of vivacious vodka cocktails from all over the world, including Black Russian, Key West Cooler and Mandarin Blossom.

COCKTAIL RECIPES

This selection includes flavoured vodkas.

ABSOLUT HERO

1 oz. (30 ml) blackcurrant vodka
1 oz. (30 ml) lemon vodka
1 oz. (30 ml) melon liqueur
⅔ oz. (20 ml) freshly squeezed lime juice
⅔ oz. (20 ml) egg white
soda water to fill
lime wedge as garnish

Shake all the ingredients, except the soda. Strain into a highball glass with ice. Fill with soda. Stir. Garnish with a lime wedge.

ACQUEDUCT

3 oz. (90 ml) vodka
½ oz. (15 ml) triple sec
½ oz. (15 ml) apricot brandy
½ oz. (15 ml) fresh lime juice

Shake all the liquids together, then strain into a martini glass and serve.

ADIOS MOTHER

1 oz. (30 ml) vodka
1 oz. (30 ml) white rum
½ oz. (15 ml) lemon juice
½ oz. (15 ml) melon liqueur
dash of triple sec
dash of blue curaçao
3 tsp. superfine (caster) sugar

Shake all ingredients and strain into a highball glass with ice.

ALEXANDER THE GREAT

2 oz. (60 ml) vodka
1 oz. (30 ml) coffee liqueur
1 oz. (30 ml) white crème de cacao
1 oz. (30 ml) light (single) cream
coffee beans as garnish
chocolate flakes as garnish

Shake all the liquid ingredients together, and strain into a martini glass. Garnish with the coffee beans and chocolate flakes, and serve.

LEFT: *Adios Mother.*

ANGLO ANGEL

1 oz. (30 ml) vodka
1 oz. (30 ml) Mandarine Napoleon brandy
1 oz. (30 ml) mandarin juice
2 dashes Angostura bitters
lime spiral as garnish

Shake all the ingredients. Strain into a cocktail glass. Add a lime spiral.

APPEASE ME

1 oz. (30 ml) vodka
1 oz. (30 ml) mango liqueur
2 oz. (60 ml) orange juice
⅔ oz. (20 ml) advocaat
1 oz. (30 ml) single cream
2 slices mango
orange slice as garnish

Blend all ingredients with ice. Pour into a highball glass. Garnish with a slice of orange and serve with a straw.

APPLE MARTINI

2 oz. (60 ml) vodka
⅔ oz. (20 ml) apple sour liqueur
⅓ oz. (10 ml) Cointreau

Shake all the ingredients. Strain into a chilled cocktail glass.

APRÈS SKI

1 oz. (30 ml) crème de menthe
½ oz. (15 ml) Pernod
½ oz. (15 ml) vodka
lemonade to fill

Pour over ice cubes in a highball glass. Top up with lemonade and serve.

AQUAMARINE

1 oz. (30 ml) vodka
2/3 oz. (20 ml) peach schnapps
⅓ oz. (10 ml) blue curaçao
⅓ oz. (10 ml) Cointreau
3 oz. (90 ml) apple juice

Shake all the ingredients. Strain into an old-fashioned glass with ice.

BELOW: *Après Ski.*

AVIATION 2

2 oz. (60 ml) vodka
1 oz. (30 ml) maraschino liqueur
⅔ oz. (20 ml) freshly squeezed lemon juice
maraschino cherry
twist of lemon as garnish

Shake all the ingredients. Strain into a cocktail glass. Drop a maraschino cherry in the drink and add a twist of lemon.

BALI TRADER

2 oz. (60 ml) vodka
⅔ oz. (20 ml) green banana liqueur
⅔ oz. (20 ml) pineapple juice

Shake all the ingredients. Strain into a cocktail glass.

BELOW: *Banana Peel.*

BALLET RUSSE

1 oz. (30 ml) Smirnoff vodka
½ oz. (15 ml) crème de cassis
½ oz. (15 ml) fresh lime juice
½ oz. (15 ml) lemon juice

Shake all ingredients with ice. Strain into a cocktail glass.

BANANA PEEL

2 oz. (60 ml) vodka
2 oz. (60 ml) banana liqueur
1 oz. (30 ml) orange juice

Shake all ingredients and strain into an old-fashioned glass.

BIKINI

2 oz. (60 ml) vodka
1 oz. (30 ml) white rum
4 oz. (129 ml) milk or single cream
½ oz. (15 ml) gomme

Shake the ingredients together, then strain into a highball glass and serve.

BLACK MAGIC

2 oz. (60 ml) vodka
1 oz. (30 ml) Kahlua
dash of lemon juice
twist of lemon to serve

Stir all the ingredients together, then pour into an ice-filled old-fashioned glass. Serve with the lemon twist.

BLACK MAGIC (ALT)

1 oz. (30 ml) vodka
⅔ oz. (20 ml) Kahlua
dash of freshly squeezed lemon juice
8 oz. (240 ml) cold coffee

Pour ingredients into an old-fashioned glass filled with ice. Stir.

BLACK MARIA

1 oz. (30 ml) vodka
1 oz. (30 ml) dark rum
4 oz. (120 ml) cold coffee

Shake all the ingredients together, then strain into an ice-filled old-fashioned glass and serve.

BLACK RUSSIAN

1 oz. (30 ml) vodka
⅔ oz. (20 ml) Kahlua

Pour the vodka, then the Kahlua into an old-fashioned glass straight up or over crushed ice, then serve.

BLOOD SHOT

1 oz. (30 ml) vodka
½ oz. (15 ml) lemon juice
2 oz. (60 ml) condensed consommé
½ oz. (15 ml) tomato soup
dash of ketchup
½ oz. (15 ml) Worcestershire sauce
celery salt as garnish
cucumber slice as garnish

Shake all ingredients and strain into a cocktail glass. Garnish with a sprinkle of celery salt and a slice of cucumber.

BLOODY CAESAR SHOOTER

1 clam
1 oz. (30 ml) vodka
1 oz. (30 ml) tomato juice
2 drops Worcestershire sauce
2 drops Tabasco
½ tsp. horseradish purée
1 pinch celery salt

Put the clam in the bottom of the shot glass, then shake the rest the ingredients together in a shaker. Strain into the glass and serve.

BELOW: *Black Russian*

It was in 1921 that Fernand "Pete" Petiot first combined tomato juice, vodka, salt, pepper and Worcestershire sauce at Harry's New York Bar in Paris. Vodka and tomato had been mixed together before in Europe (vodka was still unknown in the United States back then) but it was the addition of Worcestershire sauce that gave this drink the edge.

Hotel magnate John Astor tasted it, and asked Petiot to go to New York, to the St. Regis Hotel. Astor insisted Petiot rename this drink Red Snapper because he felt "bloody" was too much for customers to take. Petiot used gin because vodka was not available in the U.S. at that time. A customer, Prince Serge Obolensky, requested his drink be "spiced up." Petiot added the Tabasco sauce. After a time, it became known as a Bloody Mary in the U.S. too. American George Jessel was hired in the 1950s by Heublein, who had just acquired the rights to vodka, to introduce the Bloody Mary to America. He succeeded.

As with many classics, confusion reigns as to whom the cocktail is named after. Was it named after Mary I, the Tudor Queen of England who had a reputation for butchering Protestants? Or after another Mary who had caught Petiot's eye?

The use of a celery stick as a garnish originated in the 1960s at Chicago's Ambassador East Hotel. An unnamed celebrity asked for a Bloody Mary, but didn't get a swizzle stick in the glass. He grabbed a stalk of celery from the relish tray to stir his Bloody Mary and history was made.

BLOODY MARY

BELOW: *Blowout.*

1½ oz. (45 ml) vodka
5 oz. (150 ml) tomato juice
½ oz. (15 ml) fresh lemon juice
pinch celery salt
2 dashes Worcestershire sauce
2 dashes Tabasco sauce
ground black pepper
celery stick (optional)
lemon wedge as garnish

Fill a highball with ice, then pour in the tomato and lemon juices. Add the vodka. Add the spices. Stir. Add black pepper. Garnish with a lemon wedge, a stirrer and a celery stick if requested.

ABOVE: *Bloody Mary.*

BLOWOUT

1 oz. (30 ml) vodka
1 oz. (30 ml) triple sec
dash of orange juice
dash of Bacardi white rum
dash of lemon-lime soda

Shake all the ingredients, except for the lemon-lime soda, with ice and strain into a cocktail glass. Add the lemon-lime soda, stir gently, then serve.

BLUE DOVE

1 oz. (30 ml) vodka
1 oz. (30 ml) blue curaçao
lemonade to fill
whipped cream

Pour the blue curaçao and vodka over ice in a highball glass. Then stir, and top up with lemonade. Float whipped cream on the top and serve.

BLUE MANDARIN

2 oz. (60 ml) Absolut Mandarin vodka
1 oz. (30 ml) blue curaçao
3 oz. (90 ml) sour mix
lemon slice as garnish

Shake all the ingredients with ice and strain into a chilled cocktail glass. Garnish with a slice of lemon.

BLUE MARTINI

2 oz. (60 ml) vodka
⅓ oz. (10 ml) blue curaçao
⅓ oz. (10 ml) freshly squeezed lemon juice
8 fresh blueberries

Add blueberries to a shaker. Muddle. Add remaining ingredients. Shake. Strain into a cocktail glass.

BELOW: *Blue Martini.*

ABOVE: *Bongo Lemon Crush.*

BLUE ORCHID

2 oz. (60 ml) Absolut Citron vodka
dash of blue curaçao
dash of sour mix
dash of lime juice
dash of triple sec
lemon twist as garnish

Shake all the ingredients with ice. Strain into a chilled cocktail glass. Garnish with a twist of lemon.

BONGO LEMON CRUSH

3 oz. (90 ml) lemon vodka
4 lemon wedges
2 tsps caster sugar

Muddle the lemon wedges and the sugar in a mixing glass. Add the vodka to muddled mixture and stir. Strain into an old-fashioned glass filled with crushed ice.

BONZA MONZA

1 oz. (30 ml) vodka
⅔ oz. (20 ml) crème de cassis
2 oz. (60 ml) grapefruit juice

Pour ingredients into an old-fashioned glass full of crushed ice and stir.

BOSTON BULLET (ALT)

2 oz. (60 ml) chilled dry vodka
spray of dry vermouth from an atomizer
green olive stuffed with an almond to serve

Pour the vodka into a chilled martini glass with a spray of dry vermouth. Add the olive and serve.

BREAKFAST BAR

2 oz. (60 ml) vodka
handful cherry tomatoes
1 leaf fresh basil
pinch ground coriander
pinch celery salt
chopped chives
pinch ground pepper

Blend all the ingredients together, then strain into an ice-filled highball glass.

BULLSHOT

1⅔ oz. (50 ml) vodka
5 oz. (150 ml) beef bouillon
dash of freshly squeezed lemon juice
2–3 dashes Worcestershire sauce

celery salt
Tabasco sauce
black pepper

Shake bouillon, lemon juice, Tabasco and Worcestershire sauces with vodka. Strain into a highball glass full of ice cubes. Add black pepper. Serve with a stirrer.

CAESAR

2 oz. (60 ml) Absolut Pepper vodka
clamato juice
Worcestershire sauce
horseradish sauce
white pepper
celery salt
splash of fino sherry
splash of orange juice
Tabasco sauce to taste

Pour the vodka over ice in a highball glass. In a pitcher, combine the other ingredients, then add mix to the vodka and ice, and stir.

CAIPIROVSKA

2 oz. (60 ml) vodka
1 lime, diced
dash of freshly squeezed lime juice
2 tsp. caster sugar

Muddle diced lime and sugar in an old-fashioned glass. Add vodka and lime juice. Fill glass with crushed ice, stir, and serve with a straw.

LEFT: *Caipirovska.*

CAJUN MARTINI (ALT)

2 oz. (60 ml) chilled vodka
spray of dry vermouth from an atomizer
jalapeno chilli to serve

Pour the vodka into a chilled martini glass with a spray of dry vermouth. Add the chilli and serve.

CAPE CODDER

1½ oz. (45 ml) vodka
3 oz. (90 ml) cranberry juice
1 wedge of lime

Pour vodka and cranberry juice into a highball glass over ice. Stir well, add the lime and serve.

CARGO

2 oz. (60 ml) vodka
1 oz. (30 ml) white crème de menthe
2 fresh mint leaves

Rub the rim of an old-fashioned glass with one of the mint leaves. Pour the crème de menthe and vodka over ice cubes into the glass. Garnish with the other leaf and serve.

CHAMBORD KAMIKAZE

3 oz. (90 ml) vodka
½ oz. (15 ml) Cointreau
½ oz. (15 ml) lemon juice
½ oz. (15 ml) simple syrup
½ oz. (15 ml) Chambord
½ lime, sliced
lime slice to garnish

ABOVE: *Cargo.*

Place all ingredients (including the sliced lime) in a large shaker with ice. Shake violently. Strain and pour into a cocktail glass. Garnish with a slice of lime.

CHARTREUSE DRAGON

2 oz. (60 ml) vodka
2 oz. (60 ml) lychee juice
⅔ oz. (20 ml) green Chartreuse
⅓ oz. (10 ml) blue curaçao
dash of freshly squeezed lime juice
lemon-lime soda to fill

Shake all ingredients, except lemon-lime soda. Strain into a highball glass full of ice cubes. Fill with lemon and lime soda. Stir.

CHEE CHEE

1 oz. (30 ml) vodka
1 oz. (30 ml) pineapple juice
1 oz. (30 ml) coconut milk
pineapple wedge as garnish

Shake all ingredients and strain into a cocktail glass. Garnish with a pineapple wedge.

CHERRY BLOSSOM MARTINI

2 oz. (60 ml) vodka
½ oz. (15 ml) black forest cherry liqueur
½ oz. (15 ml) triple sec
dash of Rose's Lime Juice
cherry blossom as garnish

Shake all the ingredients with ice and strain into a martini glass. Garnish with a cherry blossom on top.

CHI CHI

2 oz. (60 ml) vodka
1 oz. (30 ml) coconut cream
3 oz. (90 ml) pineapple juice

Blend all the ingredients together and pour into a large goblet.

CHIQUITA

2 oz. (60 ml) vodka
⅓ oz. (10 ml) banana liqueur
⅓ oz. (10 ml) lime juice
½ banana, sliced
pinch of caster sugar

Shake all ingredients and strain into a cocktail glass. Garnish with a slice of banana.

CHOCOLATE MARTINI

2 oz. (60 ml) vodka
½ oz. (15 ml) crème de cacao

Pour ingredients into shaker filled with ice then pour into martini glass.

CHOCOLATE MINT MARTINI

2 oz. (60 ml) vodka
1 oz. (30 ml) white crème de cacao
dash of white crème de menthe

Pour ingredients into a mixing glass with ice and stir. Strain into a cocktail glass.

CLIFFHANGER

2 oz. (60 ml) pepper vodka
1 oz. (30 ml) Cointreau
⅔ oz. (20 ml) lime cordial
lime twist as garnish

Shake all the ingredients. Strain into a cocktail glass. Add a twist of lime.

COBBLER

1 tsp. icing sugar
1 oz. (30 ml) soda water
2 oz. (60 ml) vodka or any spirit
seasonal fruit as garnish

In an old-fashioned glass, dissolve the sugar in the soda water, then fill the glass with ice. Stir in the spirit, garnish with the fruit and serve.

COOL MARTINI

2 oz. (60 ml) vodka
⅔ oz. (20 ml) apple juice
½ oz. (15 ml) Cointreau
½ oz. (15 ml) freshly squeezed lemon juice

Shake all the ingredients. Strain into a cocktail glass.

COPENHAGEN

2 oz. (60 ml) vodka
½ oz. (15 ml) aquavit
slivered blanched almonds as garnish

Shake and strain into a cocktail glass and garnish with almonds.

CORDLESS SCREWDRIVER

1 oz. (30 ml) chilled vodka
1 orange wedge
sugar

Coat the orange wedge in the sugar and pour the vodka into a shot glass. Drink the vodka, then eat the orange.

COSMOPOLITAN 2

1 oz. (30 ml) vodka
½ oz. (15 ml) triple sec
½ oz. (15 ml) cranberry juice
juice of 1 lime
flamed orange peel to garnish

Shake all the ingredients, then strain into a chilled cocktail glass and serve with the flaming garnish.

ABOVE: *Cordless Screwdriver.*

CROC COOLER

1 oz. (30 ml) lemon-flavoured vodka
1 oz. (30 ml) Midori
½ oz. (15 ml) Cointreau
2 oz. (60 ml) sour mix

Pour the ingredients into a highball glass filled with ice. Stir, and serve with a straw.

CYBER PUNCH

1 oz. (30 ml) vodka
1 oz. (30 ml) gin
1 oz. (30 ml) white rum
1 oz. (30 ml) triple sec

ABOVE: *Cosmopolitan.*

½ oz. (15 ml) beer
1 oz. (30 ml) lemon juice
1 tsp. caster sugar
dash of grenadine

Pour all ingredients into a highball glass and stir. Serve with a stirrer and a straw.

CZARINA

2 oz. (60 ml) vodka
1 oz. (30 ml) apricot brandy
½ oz. (15 ml) dry vermouth
dash of Angostura bitters

Stir all the liquids together and strain into martini glass. Serve.

DEEP FREEZE

1 oz. (30 ml) peppermint schnapps
½ oz. (15 ml) vodka

Pour in a chilled shot glass or on the rocks in an old-fashioned glass. Stir. Serve.

DEMON POSSESSION

1 oz. (30 ml) citrus vodka
1 oz. (30 ml) light rum
dash of blue curaçao
lemon-lime soda
maraschino cherry as garnish

Shake all ingredients with ice. Strain into chilled highball glass filled with crushed ice and garnish with a maraschino cherry.

RIGHT: *Czarina.*

COSMOPOLITAN

Who invented the Cosmopolitan? There are several theories, one of which leads to the gay community in Provincetown, Massachusetts. The name Cheryl Cook is whispered in connection with this snazzy snifter, but no one of that name comes forward. From its success there, it went to New York (in search of fame and fortune) and became the most requested cocktail for some years.

In the past decade or so, this cocktail has become a megastar as a new-style Martini. Author Hunter S. Thompson mentioned the Cosmopolitan in an affidavit used in his legal case, The People of the State of Colorado v. Hunter S. Thompson. It gained notoriety, too, in Sex and the City as the one cocktail both good and bad girls drink.

It has become a favourite worldwide and, as such, is guaranteed a place in the history books. The best Cosmopolitans are those which succeed in balancing the sweet and sour; for instance, if not enough freshly squeezed lime juice is used it becomes too sweet. And sweet is not how a Cosmopolitan should taste. Nor should it be so tart that it makes you salivate. It should be served in a chilled cocktail glass, its sides glistening with moisture (this chill factor keeps the drink colder for longer).

Martini historian Barnaby Conrad III once described the Cosmopolitan as the "stealth Martini," and stealthy it is, having worked its way into the very fabric of our cocktail culture without too much trouble. Its critics claim it is not complex enough in flavour, but those who love it, adore it.

1½ oz. (45 ml) vodka
5 oz. (150 ml) tomato juice
½ oz. (15 ml) fresh lemon juice
pinch celery salt
2 dashes Worcestershire sauce
2 dashes Tabasco sauce
ground black pepper
celery stick (optional)
lemon wedge as garnish

Fill a highball glass with ice, then pour in the tomato and lemon juices. Add the vodka. Add the spices. Stir. Add black pepper. Garnish with a lemon wedge, a stirrer and a celery stick if requested.

DERBY PUNCH

2 oz. (60 ml) vodka
2 oz. (60 ml) gin
2 oz. (60 ml) white rum
2 oz. (60 ml) Sauza silver tequila
½ oz. (15 ml) triple sec
1 oz. (30 ml) pineapple juice
dash of cranberry juice
lemon-lime soda to fill
maraschino cherry as garnish

Shake all ingredients, except lemon-lime soda, and strain into a highball glass with ice. Top up with lemon-lime soda. Garnish with a cherry.

DETROIT MARTINI

2 oz. (60 ml) vodka
⅔ oz. (20 ml) gomme
mint leaves

Shake all the ingredients. Strain into a cocktail glass.

DIPLOMAT

2 oz. (60 ml) vodka
3 oz. (90 ml) pineapple juice
2 oz. (60 ml) Midori
½ oz. (15 ml) lemon juice

Shake all ingredients and strain into a highball glass filled with ice. Serve with a straw.

DIRTY JANE

1 oz. (30 ml) Ketel One vodka
1 oz. (30 ml) liquid from pickled green tomatoes
thin wedge of a pickled green tomato as a garnish

Pour the vodka and pickling liquid into a shaker with ice and shake. Strain into a cocktail glass and garnish with a tomato wedge.

RIGHT: *Detroit Martini.*

DOCK OF THE BAY

1 oz. (30 ml) vodka
⅔ oz. (20 ml) schnapps
3 oz. (90 ml) cranberry juice
2 oz. (60 ml) pineapple juice
lime wedge as garnish

Shake all the ingredients. Strain into a highball glass full of ice cubes. Add a lime wedge.

DOUBLE VISION

1 oz. (30 ml) lemon vodka
1 oz. (30 ml) blackcurrant vodka
4 dashes Angostura bitters
1 oz. (30 ml) apple juice

Shake all the ingredients. Strain into a cocktail glass.

DRAGON FLY

2 oz. (60 ml) vodka
⅔ oz. (20 ml) melon liqueur
dash of freshly squeezed lime juice
1 oz. (30 ml) apple juice

Shake all the ingredients. Strain into a cocktail glass.

DYEVTCHKA

1 oz. (30 ml) vodka
1 oz. (30 ml) Cointreau
½ oz. (15 ml) fresh lime juice
½ oz. (15 ml) fresh lemon juice
½ oz. (15 ml) pineapple juice
pineapple wedge as garnish
maraschino cherry as garnish

Shake with ice and strain into an old-fashioned glass. Garnish with a pineapple wedge and maraschino cherry.

EIGHT MILE CREEK COSMO

2 oz. (60 ml) Absolut Mandarin vodka
1 oz. (30 ml) fresh lime juice
1 oz. (30 ml) triple sec
1 oz. (30 ml) passionfruit purée
lemon twist as garnish

Shake with ice, strain and serve in martini glass with a lemon twist.

BELOW: *Dragon Fly.*

EVITA

1 oz. (30 ml) vodka
1 oz. (30 ml) melon liqueur
2 oz. (60 ml) freshly squeezed orange juice
1 oz. (30 ml) freshly squeezed lime juice
dash of gomme

Shake all the ingredients. Strain into an old-fashioned glass with ice.

FLIRT

1 oz. (30 ml) vodka
1 oz. (15 ml) black sambuca
2 oz. (60 ml) cranberry juice

Shake all the ingredients. Strain into a cocktail glass.

FORTUNELLA

1 oz. (30 ml) Ketel One vodka
¾ oz. (22 ml) Bombay Sapphire gin
¾ oz. (22 ml) Caravella
splash Cointreau
splash Campari
1 tsp. candied kumquat nectar
twist of lemon as garnish
twist of kumquat as garnish

Coat a shaker with the Cointreau and Campari and discard the excess. Add the remaining liquid ingredients, shake, then strain into a cocktail glass and garnish with the twists.

FORWARD PASS

1 oz. (30 ml) vodka
1 oz. (30 ml) Cointreau
1 oz. (30 ml) dry vermouth
1 oz. (30 ml) fresh lemon juice

Shake with ice and strain into a cocktail glass.

FRENCH 76

¾ oz. (22 ml) vodka
¼ oz. (8 ml) lemon juice
dash of gomme
dash of grenadine
champagne to fill

Shake the first four ingredients together, strain into a champagne flute, and top up with champagne.

FRENCH KISS 1

1 oz. (30 ml) vodka
1 oz. (30 ml) crème de mure
½ oz. (15 ml) white crème de cacao
1 oz. (30 ml) double cream

Mix the ingredients together in a shaker, then strain into a martini glass.

FRENCH MARTINI 1

2 oz. (60ml) vodka
dash of Chambord liqueur
dash of pineapple juice

Shake all the ingredients together and strain into a cocktail glass.

FRENCH MARTINI 2

1 oz. (30 ml) vodka
⅓ oz. (10 ml) Chambord liqueur
dash of pineapple juice
dash of apple cider
pinch fleur de sel
thinly sliced green apples

Line a cocktail glass with green apples, with the skins coated with fleur de sel. Shake the vodka, Chambord, and juice with ice. Chill and strain into the cocktail glass. Shake the cider until it foams and pour on top.

FRENCH MARTINI 3

3 oz. (90 ml) vodka
1 oz. (30 ml) raspberry liqueur
3 oz. (90 ml) pineapple juice
pineapple slice as garnish

Shake with ice, strain and serve in a cocktail glass. Garnish with a slice of pineapple.

FRUIT PASSION

1 oz. (30 ml) vodka
½ oz. (15 ml) passionfruit pulp
1 oz. (30 ml) rum
pineapple juice
pineapple wedge as garnish

Half-fill a highball glass with ice and pour in the vodka, rum and passion-fruit pulp. Top up with the pineapple juice. Garnish with a pineapple wedge.

FUEL FOR THOUGHT

1 oz. (30 ml) lemon vodka
⅓ oz. (10 ml) blue curaçao
dash of sour mix
dash of pineapple juice
maraschino cherry as garnish

Shake all ingredients with ice and strain into a cocktail glass. Garnish with a cherry.

BELOW: *Fuel for Thought.*

GIADA

1 oz. (30 ml) vodka
½ oz. (15 ml) Campari
½ oz. (15 ml) Galliano
dash of pineapple juice

Shake all the ingredients together, then strain into a martini glass, and serve.

GIMLET

2 oz. (60 ml) gin or vodka
1 oz. (30 ml) lime cordial
wedge of lime to garnish

Over ice, pour the gin or vodka and lime cordial into an old-fashioned glass and serve with the lime wedge.

ABOVE: *Gimlet.*

GINGER MARTINI

2 oz. (60 ml) vodka
freshly grated ginger
dash of gomme
orange zest
slice of fresh ginger as garnish

Shake with ice. Strain into a chilled cocktail glass. Garnish with a slice of fresh ginger.

GODMOTHER

2 oz. (60 ml) vodka
1 oz. (30 ml) amaretto

Pour the vodka and amaretto into an ice-filled old-fashioned glass and serve.

GOOSE BUMPS

1 oz. (30 ml) vodka
1 oz. (30 ml) red cherry purée
dash of cherry liqueur
champagne to fill

Pour vodka, cherry purée, and cherry liqueur into a champagne glass. Stir, then top up with champagne.

GREEN DINOSAUR

⅔ oz. (20 ml) vodka
⅔ oz. (20 ml) gold tequila
⅔ oz. (20 ml) light rum
⅔ oz. (20 ml) gin
⅔ oz. (20 ml) triple sec
1 oz. (30 ml) freshly squeezed lime juice
dash of gomme
dash of melon liqueur

Pour ingredients, except melon liqueur, into a shaker. Shake, then pour into a highball glass full of ice. Float the melon liqueur over the top.

GREEN TEA MARTINI

2 oz. (60 ml) lemon vodka
1½ oz. (45 ml) green tea, pre-steeped
dash of triple sec
squeeze lemon
lemon twist as garnish

Pour vodka, green tea, and triple sec into a shaker with ice. Add a squeeze of lemon and shake. Strain into a chilled cocktail glass and add a twist of lemon.

HAIR RAISER

2 oz. (60 ml) 100 proof vodka
1½ oz. (45 ml) rye whiskey
½ oz. (15 ml) fresh lemon juice

Mix together the vodka, whiskey and lemon juice in a shaker. Strain into a martini glass and serve.

HARVEY WALLBANGER

2 oz. (60 ml) vodka
5 oz. (150 ml) freshly squeezed orange juice
1 oz. (30 ml) Galliano
slice of orange

Pour vodka and orange juice into a high-ball glass full of ice and stir. Float Galliano on top. Garnish with a slice of orange and serve with a stirrer.

HOT ICE

1 oz. (30 ml) pepper vodka
½ oz. (15 ml) blue curaçao
½ oz. (15 ml) pineapple juice
lemon-lime soda to fill

Shake the first three ingredients with ice and strain into a highball glass with ice. Top with lemon-lime soda.

BELOW: *Harvey Wallbanger.*

ICEBERG

2 oz. (60ml) vodka
dash of Pernod

Stir the vodka and Pernod together, then pour into an ice-filled old-fashioned glass and serve.

IMAGINE!

2 oz. (60 ml) vodka
2 oz. (60 ml) clear apple juice
pulp of one passionfruit
ginger ale to fill

BELOW: *Iceberg.*

Scoop out the passionfruit pulp and put into a shaker with crushed ice. Add vodka and apple juice. Shake. Pour into a highball glass full of ice. Fill with ginger ale. Stir.

JAMES BOND

1 oz. (30 ml/2 tbsp.) vodka
1 sugar cube
3 dashes Angostura bitters
chilled champagne

Place the sugar cube in a champagne flute and soak in the bitters, then pour on the vodka. Fill with champagne and serve.

JOE COLLINS

2 oz. (60 ml) vodka
1 oz. (30 ml) lemon juice
1 tsp. caster sugar
dash of Angostura (optional)
soda water to fill

Place the first three ingredients in a tall glass half-filled with ice, and stir to mix. Top up with soda. Stir gently.

JOE FIZZ

2 oz. (60 ml) vodka
1 oz. (30 ml) lemon juice
1 tsp. caster sugar
dash of Angostura (optional)
soda water to fill

Shake and strain into a tall glass, and top up with soda.

JOE SOUR

  2 oz. (60ml) vodka
  ¾ oz. (22 ml) lemon juice
  ½ tsp. caster sugar

Shake and strain into a sour glass.

KAMIKAZE

  1 oz. (30 ml) vodka
  1 oz. (30 ml) freshly squeezed lime juice
  1 oz. (30 ml) triple sec

Shake all the ingredients together, then strain into a shot glass.

KEY WEST COOLER

  ½ oz. (15 ml) vodka
  ½ oz. (15 ml) peach liqueur
  ½ oz. (15 ml) Midori
  ½ oz. (15 ml) Malibu
  2 oz. (60 ml) cranberry juice
  2 oz. (60 ml) orange juice
  orange slice as garnish

Fill a highball glass with ice and add the vodka, then peach liqueur, Midori and Malibu. Add the cranberry and orange juices. Stir well. Garnish with a slice of orange.

KURRANT AFFAIR

  1 oz. (30 ml) blackcurrant vodka
  1 oz. (30 ml) lemon vodka
  4 oz. (120 ml) apple juice

Shake all the ingredients. Strain into a highball glass with ice.

LA DOLCE VITA

  1 oz. (30 ml) vodka
  5 seedless grapes
  1 tsp. honey
  Prosecco dry sparkling wine to fill

Muddle the grapes in a shaker. Add vodka and honey. Shake. Strain into a champagne glass. Fill with Prosecco.

LEFT: *James Bond.*

LIMEY

1 oz. (30 ml) lemon vodka
1 oz. (30 ml) orange liqueur
1 oz. (30 ml) freshly squeezed lime juice

Shake all the ingredients. Strain into a cocktail glass.

LONG ISLAND ICED TEA

½ oz. (15 ml/1 tbsp.) light rum
½ oz. (15 ml/1 tbsp.) vodka
½oz. (15 ml/1 tbsp.) gin
½ oz. (15 ml/1 tbsp.) tequila
½ oz. (15 ml/1 tbsp.) triple sec
juice of 1 lime
cola

Squeeze the lime into a highball, then add ice cubes and the spirits. Stir and fill up with cola. Serve with straws.

LOVE FOR SALE

1 oz. (30 ml) Absolut Mandarin vodka
½ oz. (15 ml) passionfruit liqueur
½ oz. (15 ml) pineapple juice
½ oz. (15 ml) orange juice
maraschino cherry as garnish

Shake all the ingredients with ice and strain into a chilled cocktail glass. Garnish with a maraschino cherry.

LUCY'S

1 oz. (30 ml) vodka
1 oz. (30 ml) brown crème de cacao
1 oz. (30 ml) crème de menthe

Shake the ingredients together, then strain into an old-fashioned glass and serve.

LYCHEE MARTINI

1 oz. (30 ml) vodka
⅓ oz. (10 ml) lychee liqueur
⅓ oz. (10 ml) crème de banane
1 oz. (30 ml) pineapple juice

Shake all the ingredients. Strain into a cocktail glass.

LEFT: *Long Island Iced Tea.*

MADRAS

1½ oz. (45 ml) vodka
4 oz. (120 ml) cranberry juice
1 oz. (30 ml) orange juice
lime wedge as garnish

Pour the liquid ingredients into a highball glass over ice. Add the lime wedge and serve.

MADROSKA

2 oz. (60 ml) vodka
3 oz. (90 ml) apple juice
2 oz. (60 ml) cranberry juice
1 oz. (30 ml) freshly squeezed orange juice

Pour ingredients into a highball glass filled full of ice.

ABOVE: *Mandarin Martini.*

MANDARIN BLOSSOM

3 oz. (90 ml) Absolut Mandarin vodka
3 oz. (90 ml) Midori
dash of cranberry juice
dash of orange juice
maraschino cherry as garnish
lemon slice as garnish

Pour ingredients into a shaker with ice. Shake well and strain into a highball glass filled with ice. Garnish with a cherry and a lemon slice.

MANDARINE MARTINI

1½ oz. (45 ml/3 tbsp.) gin
½ oz. (15 ml/1 tbsp.) vodka
splash Mandarine Napoleon
dash Cointreau
twist of mandarin to garnish

Pour the liqueurs in the shaker. Coat and discard surplus. Shake the spirits. Strain into a martini glass. Add the twist and serve.

MANDARIN PUNCH

2 oz. (60 ml) Absolut Mandarin vodka
dash of orange juice
dash of cranberry juice
dash of grapefruit juice
dash of cherry juice
dash of lemon-lime soda
maraschino cherry as garnish

Shake all ingredients, except lemon-lime soda, with ice. Strain into a chilled cocktail glass. Add a dash of soda. Garnish with a cherry.

MELLO JELL-O SHOT (MAKES 15 3-OZ. SHOTS)

1 pint (475 ml) Stolichnaya vodka
1½ pints (710 ml) Midori melon liqueur
8 sheets baker's gelatin
1 pint (475 ml) watermelon juice
4 oz. (120 ml) gomme syrup
½ cup dried currants (optional)
3 oz. (90 ml) crème de cassis (optional)

Heat 8 oz. (240 ml) Midori with 3 sheets gelatin until dissolved. Cool. Add the remaining Midori. For each drink, pour 1½ oz. (45 ml) into a cordial glass. Leave for 20 minutes in the freezer. Then blend 8 oz. (240 ml) of watermelon juice with the gomme syrup and 5 sheets gelatin. Heat until dissovled. Cool. Add the remaining watermelon juice and vodka. Pour 1½ oz. (45 ml) to fill each glass. Dried currants soaked in creme de cassis can be added before it sets.

MELON MARTINI

1⅔ oz. (50 ml) vodka
a quarter of a slice of watermelon
dash of freshly squeezed lemon juice

Muddle melon in a shaker. Add ice and vodka. Shake. Strain into a cocktail glass.

METRO MARTINI

1 oz. (30 ml) Stolichnaya Razberi vodka
dash of triple sec
dash of Rose's Lime Juice
½ oz. (15 ml) Chambord liqueur
juice of half a lime
lime wedge as garnish

Shake with ice. Strain into a chilled cocktail glass and garnish with a lime wedge.

METROPOLIS

2 oz. (60 ml) Absolut Mandarin vodka
½ oz. (15 ml) Mandarine Napoléon liqueur
½ oz. (15 ml) fresh lemon juice
dash of gomme syrup

Shake all ingredients with ice and strain into a cocktail glass.

METROPOLITAN 1

2 oz. (60 ml) Absolut Kurrant vodka
½ oz. (15 ml) Rose's Lime Cordial
½ oz. (15 ml) lime juice
1 oz. (30 ml) cranberry juice
lime wedge to garnish

Shake and strain into a cocktail glass and garnish with the wedge of lime.

METROPOLITAN 2

2 oz. (60 ml) blackcurrant vodka
⅔ oz. (20 ml) Cointreau
1 oz. (30 ml) cranberry juice
dash of freshly squeezed lime juice

Shake all the ingredients. Strain into a cocktail glass.

MILKSHAKE

2 oz. (60 ml) vanilla vodka
1 oz. (30 ml) Frangelico
1 oz. (30 ml) Baileys Irish Cream

Shake with ice. Strain into a chilled cocktail glass.

MISTY

1 oz. (30 ml) vodka
1 oz. (30 ml) Cointreau
1 oz. (30 ml) apricot brandy
dash of crème de banane

Stir all the ingredients together, then strain into a martini glass and serve.

MOLOTOV COCKTAIL

3 oz. (90 ml) Finlandia vodka
½ oz. (15 ml) Black Bush Irish whiskey
½ oz. (15 ml) Irish Mist

Shake and strain into cocktail glasses.

RIGHT: *Molotov Cocktail.*

When Heublein president John Martin bought the Smirnoff licence, it nearly lost him his reputation. Remember this was the 1950s, and the Soviet Union and the U.S. were hardly the best of friends. Smirnoff had been manufactured in Russia since 1820, and the Smirnoff family were masters of the art of vodka making, with a monopoly on the supply of the spirit to the royal court of Tsar Alexander III. The family's fate was less exalted after the Revolution, and the recipe for vodka escaped with the Smirnoff family to France, and eventually to America, when Martin saw the future was white and clear.

The purchase became known as Martin's folly in the trade, and he took off on a nationwide trip to promote the spirit. In Hollywood, he met with the owner of the Cock 'n' Bull, the astute Jack Morgan. Morgan had a glut of ginger beer, and a friend who was trying to sell a whole load of copper mugs. The three of them created the Moscow Mule—vodka, ginger beer and an ounce of lime juice, to be sold in a copper mug. The mug was stamped with a kicking mule, a symbol of the cocktail's kick.

The Moscow Mule caught the imagination of the younger generation, if only because its gimmickry was appealing. The effervescence of ginger beer, a tart flavour from lime juice and the smooth nontaste of alcoholic vodka ensured its success in a country keen to adopt anything new and exciting.

You can replace ginger beer with ginger ale, but it won't taste the same. But, more important, make sure you serve it cold!

1½ oz. (45 ml) vodka
½ oz. (15 ml) fresh lime juice
ginger beer to fill
lime wedge as garnish

Pour the vodka and the lime juice into a highball over ice. Fill to the top with ginger beer. Stir. Garnish with a lime wedge and serve with a stirrer.

MOTHER'S MILK

1 oz. (30 ml) vodka
½ oz. (15 ml) gin
½ oz. (15 ml) Tia Maria
½ oz. (15 ml) Orgeat syrup
4 oz. (120 ml) milk or single cream

Shake all the ingredients together, then pour into an old-fashioned glass and serve.

MUDSLIDE

2 oz. (60 ml) vodka
2 oz. (60 ml) Kahlua
2 oz. (60 ml) Baileys Irish Cream

Mix with cracked ice in a shaker. Strain and serve in a chilled highball glass.

NEW ENGLAND ICED TEA

1 oz. (30 ml) vodka
1 oz. (30 ml) triple sec/Cointreau
1 oz. (30 ml) gold tequila
1 oz. (30 ml) light rum
1 oz. (30 ml) gin
1 oz. (30 ml) freshly squeezed lime juice
1 oz. (30 ml) gomme
cranberry juice to fill

Pour ingredients, except cranberry juice, into a shaker. Shake. Strain into a highball glass with ice. Fill with cranberry juice.

ONION BREATH

2 oz. (60 ml) vodka
½ oz. (15 ml) vinegar from cocktail onions
1 drop Worcestershire sauce
½ oz. (15 ml) lemon juice
2 cocktail onions to garnish

Shake all the ingredients together, then strain into a martini glass. Serve, garnished with the onions.

ORANGE BLOSSOM

2 oz. (60 ml) vodka
2 oz. (60 ml) orange juice
dash of orange flower water

Shake all the ingredients together and strain into a cocktail glass.

RIGHT: *Onion Breath.*

ORANGE CAIPIROVSKA

2 oz. (60 ml) orange vodka
⅔ oz. (20 ml) freshly squeezed lemon juice
half an orange, diced
1 tsp. caster sugar

Muddle the orange and sugar in an old-fashioned glass. Add remaining ingredients. Fill with crushed ice.

ORANGE HUSH

3 oz. (90 ml) Absolut Mandarin vodka
1 oz. 30 ml) Grand Marnier
½ oz. (15 ml) fresh orange juice
½ oz. (15 ml) fresh lemon juice

Shake with ice and strain into a chilled cocktail glass with an icing sugar rim.

OYSTER SHOOTER (AKA HEARTSTARTER)

½ oz. (15 ml) vodka
½ oz. (15 ml) tomato juice
dash of cocktail sauce
dash of Worcestershire sauce
dash of Tabasco sauce
1 fresh oyster

Shake all the ingredients with ice and strain into a chilled cocktail glass.

PARSON'S NOSE

2 oz. (60 ml) vodka
½ oz. (15 ml) amaretto
½ oz. (15 ml) crème de peche
dash of Angostura bitters

Stir all the ingredients together, then strain into a martini glass and serve.

PEACH MOLLY

2 oz. (60 ml) vodka
dash of peach schnapps
lemon wedge as garnish
lime wedge as garnish

Shake with ice and strain into a chilled cocktail glass. Garnish with lemon and lime wedges.

BELOW: *Peach Molly.*

PEARL HARBOR

1 oz. (30 ml) vodka
⅔ oz. (20 ml) melon liqueur
1 oz. (30 ml) pineapple juice

Shake all the ingredients. Strain into a cocktail glass.

PETIT ZINC

1 oz. (30 ml) vodka
½ oz. (15 ml) Cointreau
½ oz. (15 ml) sweet vermouth
½ oz. (15 ml) orange juice
orange zest or maraschino cherry as garnish

Shake all ingredients with ice and strain into a chilled cocktail glass. Garnish with orange zest or a maraschino cherry.

BELOW: *Petit Zinc.*

PINK FETISH

1 oz. (30 ml) vodka
1 oz. (30 ml) peach schnapps
2 oz. (60 ml) cranberry juice
2 oz. (60 ml) freshly squeezed orange juice
lime wedge as garnish

Shake all the ingredients. Strain over ice into an old-fashioned glass. Add a lime wedge.

PINK LEMONADE

2 oz. (60 ml) lemon vodka
½ oz. (15 ml) Cointreau
1 oz. (30 ml) cranberry juice
juice of 1 lemon
lemon wheel as garnish

Shake with ice and serve in an old-fashioned glass with ice. Garnish with a lemon wheel.

POISON ARROW

1 oz. (30 ml) vodka
1 oz. (30 ml) light rum
2 dashes Midori
2 dashes blue curaçao
dash of pineapple juice

Shake all the ingredients with ice. Strain into chilled highball glass filled with crushed ice.

POLISH MARTINI

1½ oz. (45 ml) Wyborowa vodka
½ oz. (15 ml) Krupnik vodka
dash of apple juice

Shake and strain into a cocktail glass.

PURE BLISS

3 oz. (90 ml) lemon vodka
dash of Cointreau
dash of fresh lemon juice

Coat the rim of a chilled cocktail glass with icing sugar. Shake all the ingredients well with ice and strain into the glass.

PURPLE HOOTER

1 oz. (30 ml) citrus vodka
½ oz. (15 ml) triple sec
½ oz. (15 ml) Chambord liqueur

Shake all the ingredients together, then strain into a shot glass.

QUIET RAGE

1⅔ oz. (50 ml) vodka
2 oz. (60 ml) guava juice
2 oz. (60 ml) pineapple juice
4 fresh lychees
1 oz. (30 ml) coconut cream
dash of grenadine

Blend ingredients with crushed ice. Pour into a highball glass.

RASPBERRY MARTINI

2 oz. (60 ml) vodka
1 oz. (30 ml) crème de framboise
10 raspberries
dash of gomme

Put the gomme in a shaker and muddle the raspberries in the syrup. Add vodka and crème de framboise. Shake. Strain into a cocktail glass.

BELOW: *Purple Hooter.*

ABOVE: *Raspberry Martini.*

RED OCTOBER

2 oz. (60 ml) Stolichnaya Razberi vodka
½ oz. (15 ml) Chambord liqueur
champagne to fill
1 fresh raspberry as garnish
lemon twist as garnish

Shake first two ingredients with ice and strain into a cocktail glass. Top up with champagne and garnish with a fresh raspberry and lemon twist.

RED RUSSIAN

1 oz. (30 ml) vodka
1 oz. (30 ml) white crème de cacao
2 dashes grenadine

Shake all the ingredients together, then strain into an ice-filled old-fashioned glass and serve.

REGGAE SUNDASH

1 oz. (30 ml) lemon vodka
½ oz. (15 ml) coconut rum
6 oz. (170 ml) orange juice
dash of grenadine
dash of soda water

Shake all the ingredients, except soda, with ice and strain into a highball glass filled with ice. Serve with a straw.

ROAD RUNNER

1 oz. (30 ml) vodka
⅔ oz. (20 ml) amaretto
⅔ oz. (20 ml) coconut cream

Shake all the ingredients together, then strain into a martini glass and serve.

ROYAL BLUSH

1 oz. (30 ml) vodka
1 oz. (30 ml) crème de framboise
1 oz. (30 ml) double cream
2 dashes grenadine

Mix the ingredients together in a shaker, then strain into a martini glass and serve.

RUSSIAN BEAR

1 oz. (30 ml) vodka
1½ oz. (45 ml) single cream
¼ oz. (8 ml) crème de cacao
1 tsp. caster sugar

Shake and strain into a cocktail glass.

BELOW: *Russian Bear.*

RUSSIAN COCKTAIL

1 oz. (30 ml) vodka
1 oz. (30 ml) gin
1 oz. (30 ml) white crème de cacao

Shake all the ingredients together, then strain into a martini glass and serve.

SALTY DOG

2 oz. (60 ml) vodka
2 oz. (60 ml) grapefruit juice

Mix the vodka with the grapefruit juice in a shaker, then strain into a martini glass and serve.

ABOVE: *Screwdriver.*

SCREWDRIVER

2 oz. (60 ml) vodka
5 oz. (150 ml) freshly squeezed orange juice

Pour the vodka into a highball glass with ice. Add orange juice, stir, and serve with a stirrer.

SCROPPINO

1 oz. (30 ml) Absolut Citron vodka
6 oz. (170 ml) lemon sorbet
champagne
lemon slice (thin) as garnish

Pour all ingredients into a mixing glass and blend with a hand blender. Strain into a cocktail glass and garnish with a thin slice of lemon.

ABOVE: *Salty Dog.*

SEA BREEZE

  2 oz. (60 ml) vodka
  3 oz. (90 ml) cranberry juice
  2 oz. (60 ml) fresh grapefruit juice

Pour ingredients over ice into a highball glass. Stir and serve with a stirrer.

SEA HORSE

  1½ oz. (45 ml) vodka
  dash of Pernod
  1 oz. (30 ml) apple juice
  1 oz. (30 ml) cranberry juice
  ¼ lime, freshly squeezed
  sprig of mint as garnish

Pour ingredients into a highball glass filled with ice. Garnish with a sprig of mint.

SILVER BULLET

  2 oz. (60 ml) vodka
  1 oz. (30 ml) Kummel

Pour the vodka and Kummel into an old-fashioned glass and serve.

STUDIO 54 KISS

  3 oz. (90 ml) Ketel One vodka
  ½ oz. (15 ml) pineapple juice
  ½ oz. (15 ml) peach purée
  dash of peach liqueur
  champagne to fill
  peach wedge as garnish
  2 pineapple leaves as garnish

Combine first four ingredients in a shaker with ice. Shake well and strain into a chilled cocktail glass. Top up with champagne. Garnish with a peach wedge and two neat pineapple leaves.

STRAWBERRY CREAM TEA

  1 oz. (30 ml/2 tbsp.) Kahlua
  1 oz. (30 ml/2 tbsp.) Baileys
  1 oz. (30 ml/2 tbsp.) fraise
  1 oz. (30 ml/2 tbsp.) vodka
  1 oz. (30 ml/2 tbsp.) lassi (Indian
  yogurt drink)
  strawberry to garnish

Blend the ingredients, then pour into an ice-filled highball glass. Serve with a strawberry on the rim. Lassi gives this cocktail a lighter, cleaner flavour.

BELOW: *Sea Breeze.*

SUGAR COCKTAIL

1½ oz. (45 ml) vodka
½ oz. (15 ml) peach schnapps
dash of peach nectar
dash of sour mix

Shake with ice, and strain in a sugar-rimmed martini glass.

TABLATINI A FUSION FOR 4 SERVINGS

8 oz. (240 ml) lemon-flavoured vodka
1 pint (475 ml) pineapple juice
3 oz. (90 ml) fresh lemon juice
6 coarsely chopped stalks fresh lemon grass

Simmer the pineapple juice and lemon grass in a saucepan for 15 minutes. Take off the heat and cool. Strain into a jar and place in the refrigerator. To a pitcher filled with ice, add the pineapple juice, vodka and fresh lemon juice. Mix well. Strain into four chilled cocktail glasses.

THAT STINGER

2 oz. (60 ml) vodka
1 oz. (30 ml) white crème de menthe
chocolate mint candy

In a chilled mixing glass, stir the vodka and creme de menthe. Garnish with the mint candy.

THE ABYSS

1 oz. (30 ml) vodka
½ oz. (15 ml) white rum

½ oz. (15 ml) blue curaçao
pineapple juice

Pour the blue curaçao into a chilled cocktail glass. Combine the vodka, rum, and pineapple juice in a blender with ice. Strain into the cocktail glass over blue curaçao.

THE VACATION

2 oz. (60 ml) vanilla vodka
2 oz. (60 ml) light rum
dash of pineapple juice
dash of sour mix

Shake all the ingredients with ice. Strain into a chilled cocktail glass.

TRANSFUSION

2 oz. (60 ml) vodka
½ oz. (15 ml) pure grape juice
soda water to fill
lemon twist as garnish

Shake first two ingredients with ice, strain into an old-fashioned glass with ice. Top up with soda and garnish with a lemon twist.

VELVET HAMMER

2 oz. (60 ml) vodka
1 oz. (30 ml) white crème de cacao
1 oz. (30 ml) double cream

Mix the ingredients together in a shaker, then strain into a martini glass and serve.

RIGHT: *The Abyss.*

VERY CHANILLA (CHRIS EDWARDES)

1 oz. (30 ml/2 tbsp.) vanilla vodka
½ oz. (15 ml/1 tbsp.) cherry schnapps
1 oz. (30 ml/2 tbsp.) cherry purée
juice of 1 lime
½ oz. (15 ml/1 tbsp.) gomme syrup
4 griottine cherries

Shake the ingredients, then pour into an old-fashioned glass and serve.

VESPER

1 oz. (30 ml) Stolichnaya Gold vodka
1 oz. (30 ml) Beefeater dry gin
dash of white Lillet
lemon twist as garnish

Vigorously shake ingredients with ice and strain into a chilled cocktail glass. Garnish with the twist.

ABOVE: *Vodka Martini.*

VODKA MARTINI

2 oz. (60 ml) chilled vodka
spray of dry vermouth from an atomizer
green olive or twist of lemon to serve

Pour the vodka into a chilled martini glass with a spray of dry vermouth. Add the olive or lemon and serve.

LEFT: *Very Chanilla.*

YELLOW FEVER

....

   2 oz. (60 ml/4 tbsp.) vodka
   ⅔ oz. (20 ml/1⅓ tbsp.) Galliano
   ⅔ oz. (20 ml/1⅓ tbsp.) fresh lime juice
   1 oz. (30 ml/2 tbsp.) pineapple juice

Shake the ingredients, then strain into
a cocktail glass and serve.

WHITE RUSSIAN

....

   1 oz. (30 ml/2 tbsp.) vodka
   1 oz. (30 ml/2 tbsp.) Kahlua
   1 oz. (30 ml/2 tbsp.) heavy (double) cream

Shake the ingredients, then strain into a
martini glass and serve.

Alternatively, layer the ingredients in
an ice-filled old-fashioned glass.

COCKTAIL
CULTURE

———◆—✕◆✕—◆———

"There are only two absolutes in life: friends and vodka.
And the best times usually involve both."

Unknown

THE HISTORY

Who invented the cocktail? Answer: The first person who decided to find out what two or three different ingredients would taste like when they were mixed together. Cocktails are a product of experiment, the result of satisfied curiosity. Their fortunes have ebbed and flowed as people's tastes have changed. Some have lasted the distance, becoming legendary concoctions, others have fallen by the wayside.

Cocktails mirror the trends of their time; they are a barometer of society. They range from aristocratic concoctions to drinks of the people. They were the tipple for the "Bright Young Things" of the 1920s; the fuel that spurred Hemingway and Ian Fleming; that sharpened the wit of Dorothy Parker; that gave the disco era added froth and sparkle. Today they are cool, ironic and decidedly post-modern in attitude.

Their recipes have sealed friendships, but mostly cocktails are a matter of heated debate. No two people can agree about when they were first made, where the name came from, or even how to make them. Beware, you are entering a minefield.

The rum punches drunk in England by Boswell were, in their own way, ur-cocktails, but the term first appeared in an American dictionary in 1806, meaning "a mixed drink of any spirit bitters and sugar".

That said, no one can agree where the name came from. Some argue it came from a horse-breeding term, referring to a horse that was part thoroughbred and known as a "cocktail" because its tail was docked. Others have a fanciful tall tale that it was the name of an Aztec princess called Xochitl.

James Fenimore Cooper preferred the convoluted history of an inn-keeper called Betsy Flanagan, who stole her neighbour's chickens and served them to some of Lafayette's volunteers. She tied the tail feathers to mugs of drink and the French soldiers raised a toast of "Vive le Cock-tail!" The most plausible, if rather prosaic, explanation remains the French term coquetel, meaning a mixed drink. The fact that other tales point to New Orleans using the term early in its life gives a hint that this might be the right one.

Many countries have contributed to the evolution of the cocktail, but America remains its spiritual home – the cocktail

OPPOSITE: *Cocktail bar sophistication in the Twenties.*

ESSAYAGE-BAR

Un couturier très parisien vient d'installer, attenant
à ses salons d'essayage, un bar dernier cri.

— BARMAN, UN TROIS-PIÈCES COCKTAILS !

Dessin de BONNOTTE.

bar is the high temple, the barman a priest, the drinkers acolytes, genuflecting before the raised shaker.

The first high priest of the cult was Professor Jerry Thomas, a relentless showman and self-publicist, the Barnum of the bar, and his 1862 book *The Bar-Tender's Guide and Bon Vivant's Companion* contains the first recorded recipes of the craze that was beginning to grip America.

Thomas toured Europe, lugging £1,000-worth of silver bar equipment with him, showing off his dazzling creations such as the Blue Blazer, which would involve tossing a stream of flaming whiskey from mixing flask to mixing flask.

Cocktails have swung from sweet to dry throughout their life. In the early days, sweet was in. The main base spirit was gin, but the most common style at that time was the sweet Old Tom, rather than the dry gins that were being made by the end of the nineteenth century. Bitters were also an integral part of all the early mixes – orange, Angostura and Peychaud all feature heavily in the recipes of the time, as do the sweet liqueurs that were widely drunk.

So when the good professor invented the Martinez, it wasn't the Martini we know today, but a sweet mix of Old Tom, sweet vermouth, maraschino liqueur and bitters.

The best gauge of how the cocktail continued its passage into the heart of American social life can be charted by the number of books that appeared on the subject. Thomas's greatest rival was Harry Johnson, who in 1882 published *New And Improved Illustrated Bartender's Manual, Or How to Mix Drinks of the Present Style*, with hundreds of recipes in it. Many – most of them, even – are forgotten. Barmen have taken the ones they wanted and modified them, while ingredients and tastes have changed. Shed a tear then for the Goat's Delight, Bishop's Poker and the Hoptoad.

By the time Johnson was penning his book, the swanky hotels of the day had caught on to the craze and become cocktail laboratories. New York's Waldorf-Astoria (home of The Bronx) was typical of its day in having a cocktail created for it. In fact, the Astoria was one of the drinks edging us closer to the Martini we know today, but it has languished in the shadows as other, newer drinks have taken centre stage.

American bars had begun to spring up in the fanciest European capitals, and new creations were soon spilling out of them, like The Sidecar which was invented in Paris in 1911. In London, the Ritz, the Savoy and the RAC Club all chipped in with their own contributions to the fast-growing lexicon.

The new dry gins were at the centre of the majority of the most successful, and by bringing gin to a wider audience, cocktails were the making of the spirit as an international favourite. Gin, recognized as one of the great mixers, had found its role, and without vodka to challenge it, had the field to itself.

Despite this, gin's homeland remained a tad aloof from the phenomenon. There were cocktail bars in Britain, but it was not appropriate behaviour to be seen in one. The country had to wait until the end of the First World War for cocktails to become truly popular.

BELOW: *Sidecar.*

The end of the war brought an eruption of relief. The corsets of the Edwardian era were removed and young people, for the first time in Britain's history, began to rebel. This coincided with America starting its long experiment with banning alcohol, and so the focus of attention switched to Cuba, London, Paris, Venice and Berlin. Cocktails, however, remained quintessentially American. This was a time when the rest of the world was first bombarded with images of America. It was the birth of the movies, the first great explosion of jazz. Cocktails were bound up in this American spirit, and Europeans recovering from war drank them because they wanted to taste some of that sense of optimism and youth.

That said, in Britain cocktails remained slightly decadent. They were drunk by a small number of young middle-class artists and students. Cocktail parties weren't held in the working-class slums. Then the Second World War came and put paid to all of that frivolity.

In Cuba, meanwhile, a revolution was brewing. Cocktails had been made since the turn of the century when the Mojito, the Daiquiri and the Cuba Libre were all invented. During Prohibition, though, America decanted itself into Old Havana – literally, in the case of a Boston barman called Donovan, who transported his entire bar to the city Havana. And with the gangsters, movie stars and tourists came barmen, ready to learn from the recently formed Cuban Cantineros Club, an association and a school for cocktail makers.

Havana's hotels and bars were crammed with thirsty, partying Americans and the barmen responded. New recipes, like the Jai-Alai, the Ron Fizz, the Santiago, the Presidente, the Mary Pickford and the Caruso flooded out of the Inglaterra, the Pasaje and, most famous of all, el Floridita, home to the greatest barman of his day, Constante Ribailagua – the inventor of the frozen Daiquiri. "When it comes to cocktails, Cuba is ahead of us all. American, French and English barmen could learn a lot here," wrote Albert Crocket in the *Old Waldorf Astoria Bar Book* in 1935.

OPPOSITE: *The French Casino in New York.*

That isn't to say that America had taken the pledge. In fact, drinking increased during Prohibition and, though underground, the cocktail continued. The reason that so many new recipes were created during this period is conceivably because the taste of the bathtub gin and vodka that was being produced desperately needed to be masked, so fruit juices, other spirits and bitters were added. Long drinks such as the Screwdriver were born around this time, for this very reason.

Those who could get their hands on good-quality English gin were beginning to develop a taste for drier drinks. It is around this time that the Martini first begins to move from the 2:1 ratio of gin to vermouth to the drier style we know today.

After Repeal, America picked up where it had (officially) left off before Prohibition. Cocktails grew even through the Depression, the rationale perhaps being that it made more sense to have one strong drink than a lot of little ones.

If gin had had it all its own way before the Second World War, then the arrival of white whiskey (as vodka was first called) was soon to change all that. Over the years, vodka has outstripped its rival, muscling in on the Martini and dragging the title of hangover cure away from the Red Snapper made with gin and tomato juice to the Bloody Mary.

Some see vodka's rise as a triumph of no taste over flavour. It's certainly true that the vodkas that arrived in the United States and Britain after the war were miles away from the real stuff produced in Russia and Poland, but this flavourless base spirit allowed people to enjoy mixed drinks all the more. Vodka was new, it was light, it was cool. People have always looked for light flavours in cocktails. Even when the cream-laden drinks of the 1970s arrived, it was a virtually neutral base that drinkers wanted. Brown spirits, most notably bourbon, may have been the base for many classics, but light remains most people's main criterion for a great cocktail. Vodka was the lightest of them all.

In the 1960s cocktails hit a wall. They had become an acceptable part of society, the businessman's drink, the politician's. In the 1960s, if you wanted to be rebellious you wouldn't sip a Martini – you'd let your freak flag fly and drop acid instead.

Cocktails remained quiet until disco arrived in the late 1970s and everything went silly again. It was all mirror balls, creamy drinks, spangly tops and Gloria Gaynor. The trendiest drinks were brightly coloured and had drag-queen names like Brandy Alexander and Pink Pussy.

In Britain in the late 1970s and early 1980s there was a mini-revival when post-punk people began listening to acid jazz and salsa and took to sipping classic cocktails, but it was always more of a pose than any great love of what they were actually drinking. At the same time, though, some drinks – the Piña Colada, the Black Russian, the Tequila Sunrise and the Margarita – had escaped from the cocktail bar and made their way into pubs and clubs. The Wham! era of blonde highlights, tight shorts and fake suntans was defined by the irresistible rise of the Piña Colada.

In America, meanwhile, it all went badly wrong. It's Jimmy Carter who gets the blame for the downfall, with his speech complaining about the three-Martini lunch, although he was actually complaining about the tax-deductible lunch, not the Martinis!

BELOW: *Cocktails at the beach.*

There again, it's difficult to think of the Georgia peanut farmer being like FDR and mixing Dirty Martinis in the Oval Office every night.

This was the trigger for the Moral Majority to ride out and lynch anyone who dared to suggest drinking might not be all that anti-social. In time, cocktails became ensnared in the War Against Drugs and, as they did and people began to be scared of drinking them, so barmen tried to out-gross each other to get some attention. The names (and the drinks) got sillier – the Blow Job being a good example: a B-52 with whipped cream on the top that had to be drunk without touching the glass with your hands. Try it, and you'll see why it gets its name... If you weren't drinking dumb drinks, you were trying to pile as much alcohol in the glass as possible. Enter the Long Island Iced Tea.

Gary and Mardee Regan, in *New Classic Cocktails*, their essential guide to today's cutting-edge concoctions, define them as punk cocktails (that's the American cartoon version of punk, not the nihilistic UK one) but go on to say "without punk cocktails as an impetus we might never have seen some of these new classics... Punk cocktails smartly slapped the classicist bartenders across the face and screamed 'create, goddamn it, create'".

Then, in the mid-90s the classicists hit back. The Martini returned, with a new post-modern ironic twist along with the slice of lemon. They seem to be nostalgic for the past but, by being knowingly nostalgic, they are also detached and ironic. I remain unconvinced about much of the Martini revival, which has now reached ludicrous proportions. Gin and vodka can make Martinis, but no other spirit can be used. If you want to make a tequila "Martini", go and find another name.

This isn't some purist, fundamentalist position. There are some that work brilliantly and, anyway, the whole nature of cocktails is that they change and shift with each generation.

Progress is to be encouraged, but what seems to be happening is that in the rush to create something new, many have forgotten the basic principle of any cocktail, especially a Martini, which is that the drink highlights its base spirit while being subtly enhanced by its flavourings. There's the art of mixing a cocktail. It's about putting a new spin on perfection. A haiku in its mixing, satori when it is sipped.

The post-modern cocktail has come about as a result of there now being a greater range of high-quality spirits than ever before. The consumer and the barman are spoiled for choice. There's reposado and anejo tequila and fine mezcal, top-quality bourbon, premium gins and vodkas, and great flavoured vodka – and by that I mean the real Polish and Russian varieties, not the ersatz confected cheapies. All of this opens up a massive new range of possibilities.

It's wonderful to see this backlash against the morality of the late 1980s and early 1990s. It might not be two-fisted drinking, but people are sick of being told that everything is bad for them. Generation X wants to enjoy itself and, as we have seen in every chapter of this book, the desire for quality and flavour is what is driving the new push for the top end of the market. Cocktails are another manifestation of this – you can't make a great cocktail with poor-quality ingredients.

Why has it happened? Maybe it is simply because we just feel better about ourselves. After all, cocktails are drinks of celebration. Enjoy.

BELOW: *Cocktails in the 21st century.*

BARTENDER'S GUIDE

Who invented the cocktail? Answer: The first person who decided to find out what two or three different ingredients would taste like when they were mixed together. Cocktails are a product of experiment, the result of satisfied curiosity. Their fortunes have ebbed and flowed as people's tastes have changed. Some have lasted the distance, becoming legendary concoctions, others have fallen by the wayside.

HOW TO MAKE A COCKTAIL

It's been said many times before, but it's worth repeating that a cocktail is a combination of three things:

> **The base spirit, which gives the cocktail its main flavour and identity.**

> **The modifier/mixer, which melds with the base spirit and transports it on to another plane, but doesn't dominate it.**

> **The flavouring. This is the smallest element, but acts like the tiny detail that sets good clothes apart. It's the drop of bitters, the splash of coloured liqueur, the squeeze of a twist of lemon.**

> **Understand that, and you're on your way. Now the technical bit.**

SHAKING

First, an explanation of why you shake. It is to mix and also to chill the drink down, and give it a slight dilution, which helps to release flavours. The shaker, which should be made of stainless steel or glass, should never be more than half-filled with ice. Shaking actually doesn't take too long. A vigorous shake for 10 seconds will be ideal for the majority of cocktails. A simple indication is that the outside of the shaker should be freezing to the touch. If you shake for too long you'll end up diluting the drink. The contents are then strained though the strainer so that none of the ice ends up in the drink. Only use fresh ice cubes in the glass.

STIRRING

There's as much written about this as any part of cocktail-making. In general, stirring is used to marry flavours that go together easily, and to prevent the cocktail becoming cloudy. Whether you shake or stir your Martini is entirely up to you. As in shaking, this is another way to chill a drink down quickly. Half-fill the stirring glass with ice and agitate (15–25 seconds should do it). Normally, stirred drinks are strained into the cocktail glass, although a few are stirred in their final resting place.

BLENDING

Here, ice is whizzed up with the spirit and served unstrained, although the pile of mush that is usually served up as a frozen cocktail rather ruins the point of the drink in the first place, as all you do is freeze your insides and not taste anything. The blender is also handy for zapping fruit for Daiquiris.

MUDDLING

This means gently pressing and mixing some ingredients in the bottom of a glass, either with a pestle or the back of a spoon. It doesn't involve pulverizing them, just making a rough purée.

LAYERING

The heaviest part of a layered drink goes in first; the successive layers are gently poured in over the back of a spoon.

For some reason, this ingredient is rarely given a second thought. But water has a flavour too – and chlorinated water has a flavour that is enough to turn a potentially great cocktail into a hideous undrinkable mess. If the tap water tastes of chemicals, don't use it for ice. Use filtered water or, better still, bottled water. Ice cubes are normally used in shaking and stirring. Crushed ice is colder, but it melts and leads to quicker dilution. Freezing tonic water works, or you can make vermouth-flavoured ice cubes for the Martini.

CHILLING GLASSES

Most cocktails call for chilled glasses. The simple solution is to keep them in the freezer, but if you don't have space they can be chilled down quickly by filling them with ice and water and leaving them to stand. When you are ready to use them, shake them so the freezing water chills the outer surface, then tip all the ice and water out. Try not to pick them up by the bowl, but by the stem.

SALTING THE RIM

Don't bury the glass rim down in a pile of salt or sugar. The intention here is to coat the outside surface of the glass, not the inside. Moisten with lemon or lime juice, then carefully turn the glass side on in a saucer of the salt/sugar, or sprinkle the coating on to the glass while rotating it (make sure you have some paper to catch the mess).

FRUIT

Always use fresh, washed fruit. One tip is to roll limes before cutting them or using them for Caipirinhas. This allows them to start releasing their juices.

HOW TO CUT AND USE A TWIST

Pare small strips from a lemon, ensuring that some white pith is still attached. Holding the peel between your thumb and forefinger, give it a quick twist so that some oil sprays from the skin on to the drink. Run the twist round the rim of the glass and drop in. Longer strips can be tied into a knot and dropped in.

MEASUREMENTS

All the recipes are in the standard accepted US ounces. The Imperial British ounce is slightly smaller, but makes virtually no difference. If you are making cocktails at home, a US ounce is 2.8cl, an Imperial 2.5cl. For home mixing, an Imperial ounce is the equivalent of two-and-a-half tablespoons or six teaspoons.

It's worth getting a set of stainless-steel measuring spoons for teaspoons (tsp) and tablespoons (tbs). With practice, you'll end up knowing what an ounce looks like.

A dash is just what it says: the merest splash. When small additions are indicated (such as with maraschino), it's just a bit more than a dash.

Following a cocktail recipe is much the same as using a cookery book; you don't have to be restricted by the instructions. If you prefer more lemon juice, less (or more) alcohol, a different garnish, then go ahead and satisfy your own palate. Once you have mastered the basics, the world is your oyster (or bar).

GLASS SHAPES

1.

4.

5.

6.

1. *cocktail glass (usually for Daiquiris and Margaritas)*

2. *large highball glass*

3. *cocktail (or martini) glass*

4. *balloon/punch glass*

5. *small highball glass*

6. *collins glass*

7. *flute*

7.

VODKA
PLACES

COCKTAIL BARS

CANADA

Alberta
THE TREASURY VODKA BAR & EATERY
10004 Jasper Avenue, #100, Edmonton,
AB T5J 1R3

Ontario
PRAVDA VODKA HOUSE
44 Wellington St E, Toronto, ON M5E 1C7

Proof, the Vodka Bar
220 Bloor Street W, Toronto, ON M5S 1T8

USA

California
BAR LUBITSCH
7702 Santa Monica Boulevard, West Hollywood, CA 90046

Georgia
CZAR ICE BAR
56 E Andrews Dr NW Suite 20, Atlanta,
GA 30305

Minnesota
ST PETERSBURG RESTAURANT AND VODKA BAR
3610 France Ave N, Robbinsdale, MN
55422

Missouri
SUB ZERO VODKA BAR
308 N Euclid Ave, St. Louis, MO 63108

Nevada
RED SQUARE RESTAURANT & LOUNGE
3950 Las Vegas Fwy, Las Vegas, NV 89119

New York
PRAVDA
281 Lafayette Street, NY10012

RUSSIAN VODKA ROOM
265 W 52nd Street, NY10019

Washington D.C.
RUSSIA HOUSE RESTAURANT AND LOUNGE
1800 Connecticut Ave NW, Washington,
DC 20008

UK

Belfast
BAR RED
Ten Square Hotel, 10 Donegall Square S,
Belfast BT1 5JD

Edinburgh
REVOLUTION BAR
30A Chambers Street, Edinburgh EH1 1HU

SLIGHHOUSE
54 George IV Bridge, Edinburgh EH1 1EJ

Glasgow
REVOLUTION BAR
67-69 Renfield Street, Glasgow G2 1LF

VODKA WODKA
31–35 Ashton Lane, Glasgow G12 8SJ

London
AMERICAN BAR
The Savoy Hotel, Strand, London WC2R
0EU

BAR POLSKI
11 Little Turnstile, Holborn, London WC1V
7DX

DOOST
305–307 Kennington Road, London SE11
4QE

LAB BAR
12 Old Compton Street, London W1D 4TQ

LONG BAR
The Sanderson Hotel, 50 Berners Street,
London W1T 3NG

QUO VADIS
26–29 Dean Street, London W1D 3LL

REVOLUTION BAR
140–144 Leadenhall, London EC3V 4QT

REVOLUTION BAR
95–97 Clapham High Street, London SW4
7TB

THE TOWNHOUSE
31 Beauchamp Place, London SW3 1NU

Manchester
DRY BAR
28–30 Oldham Street, Manchester M1 1JN

KRO BAR
325 Oxford Road, Manchester M13 9PG

NICO CENTRAL BAR
Mount Street, Manchester M60 2DS

RESTAURANT BAR AND GRILL
14 John Dalton Street, Manchester M2 6JR

THE RIVER BAR
The Lowry Hotel, 50 Dearmans Place,
Salford M3 5LH

TRIBECA BAR
50 Sackville Street, Manchester M1 3WF

EUROPE
Amsterdam
CIEL BLEU BAR
Ferdinand Bolstraat 333, 1072 LH
Amsterdam

Berlin
HUDSON BAR
Elßholzstraße 10, 10781 Berlin

Madrid
MUSEO CHICOTE
Gran Vía, 12, 28013 Madrid

Paris
THE HEMINGWAY BAR
The Ritz Paris, 15 Place Vendôme, Paris
75001

LE PLAZA ATHENÉE
25 Avenue Montaigne, Paris 75008

Stockholm
EKEN BAR
Guldgränd 8, 104 65 Stockholm

DISTILLERY TOURS

SWEDEN
ABSOLUT INFO CENTER
Köpmannagatan 23, Åhus
(Offers guided tours of the Absolut Distillery
during the summer)

UK

CHASE DISTILLERY
Rosemaund Farm, Hereford HR1 3PG

SIPSMITH DISTILLERY
83 Cranbrook Road, Chiswick, London
W4 2LJ

INDEX

PICTURE CREDITS

The publishers would like to thank the following sources for their kind permission to reproduce the pictures in this book.

Advertising Archives: 15, 33, 60; **Alamy:** /Jeffrey Blackler: 68; /Mark Fagelson: 86; /Philip J Hill: 96; /Ken Howard: 95; /Chris Howes/Wild Places Photography: 77; /i food and drink: 75C, 105; /kpfoto: 36, 79; /Mary Evans Picture Library: 163; /PSL Images: 101L; /Emiliano Rodriguez: 101R; /Oren Shalev/PhotoStock-Israel: 180; /Stars and Stripes: 102-103; /Sudres/StockFood GmbH: 76; /Gary Vogelmann: 104; /Rob Wilkinson: 100; **Buffalo Trace Distillery:** 98L; **Gerrit Buntrock:** 118-119, 145; **Carlton Books:** /Karl Adamson: 111, 114, 115, 116, 117, 120BL, 120TR, 121BL, 121TR, 123, 124, 126BL, 126TR, 127, 128, 130, 131, 133, 134, 135, 136, 137, 138, 140, 142, 143, 144, 146, 147, 148, 149, 150, 151, 152BL, 153, 155, 156BL, 156TR, 174, 182, 183; **Corbis:** /Dean Conger: 42-43; /DiMaggio/Kalish: 178-179; /Everett Collection: 19; /Rose Hartman: 80-81; Hulton-Deutsch Collection: 51; /James Marshall: 22-23, 28-29; /Brian Vikander: 57; **e.t. Archive:** 50; **Getty Images:** /Sergi Alexander: 66-67; /Malcolm Ali/WireImage: 71; /Neilson Barnard: 74, 87; /Craig Barritt: 72-73; /Donald Bowers: 93; /Gael Branchereau/AFP: 54; /Charley Gallay: 99, 106; /Matthias Hoffmann: 165; /Linus Hook/Bloomberg: 58-59; /Tasos Katopodis: 64-65; /Keystone-France/Gamma-Keystone: 166; /Maurice Rougemont/Gamma-Rapho: 46; /Andrey Rudakov/Bloomberg: 82-83; /Bartek Sadowski/Bloomberg: 20; /Andrew H Walker: 92; /Alexander Zemlianichenko Jr/Bloomberg: 84; **Ketel One:** 98R, 107; **Lancut Distillery Museum:** 27, 41; **Lucas Bols:** 75L; **Panos Pictures:** /Białowieza: 24; /J C Tordai: 49; /Tyrone Wheatcroft: 52-53; **Pernod Ricard:** 75R; **Plodimex UK Ltd:** 13; **Polish Vodka Web Site:** 32, 38, 44; **Private Collection:** /Sakki: 90; /Startraks Photo: 97; **Roust:** 85; **The Seagram Co. Ltd:** 35; **Shutterstock:** 6, 8-9, 62, 88, 108, 112-113, 129, 139, 152TR, 157, 158-159, 160, 169, 171, 172, 175, 176, 177, 181, 184, 190; **United Distillers and Vintners:** 10, 16-17; **Waldemar Behn:** 91; **Williams Chase:** 94; **Wodka Restaurants Ltd:** 30, 89; **Zielona Gora Distillery:** 12

Every effort has been made to acknowledge correctly and contact the source and/or copyright holder of each picture and Carlton Books Limited apologises for any unintentional errors or omissions that will be corrected in future editions of this book.